Implementing Energy Subsidy Reforms

Implementing Energy Subsidy Reforms

Evidence from Developing Countries

Maria Vagliasindi

THE WORLD BANK
Washington, D.C.

ISBN (paper): 978-0-8213-9561-5
ISBN (electronic): 978-0-8213-9562-2
DOI: 10.1596/978-0-8213-9561-5

Cover photo: iStockphoto

Library of Congress Cataloging-in-Publication Data
Vagliasindi, Maria.
   Implementing energy subsidy reforms : evidence from developing countries / Maria Vagliasindi.
     p. cm.
   ISBN 978-0-8213-9561-5 — ISBN 978-0-8213-9562-2 (electronic)
1. Energy consumption—Developing countries. 2. Poor—Energy assistance—Developing countries. 3. Energy policy—Developing countries. I. Title.
   HD9502.D442V34 2012
   333.791'58091724—dc23

2012022848

Contents

Figures

Tables

Acknowledgments

The report was developed by a team led by Maria Vagliasindi, Lead Economist, with input from Besnik Hyseni, Nicolas Jost, Sebastian Lopez Azumendi, and Evgenia Shumilkina, all from the Energy Unit of the Sustainable Energy Department.

We would like to thank Vivien Foster, Lead Economist, Sustainable Development Network, Africa Region, and Ruslan Yemtsov, Lead Economist, Human Development Network, Social Protection Team, and other reviewers including Mamta Murthi, Acting Sector Director; Benu Bidani, Human Development Sector Unit, Europe and Central Asia Region (ECA); Caterina Ruggeri, Poverty Reduction/Economic Management, ECA; Eduardo Ley, Lead Economist, Economic Policy and Debt Department, Poverty Reduction and Economic Management Network (PREM); Giovanna Prennushi, Ulrich Bartsch, and Rinku Murgai, Economic Policy and Poverty Sector, PREM, South Asia Region (SAR); Husam Beides, Senior Energy Specialist, Energy Unit, Sustainable Development Network; Gary Stuggins, Lead Energy Economist, ECA Energy Unit; and Jon Strand, Senior Economist, and Mike Toman, Lead Economist and Manager, Environment and Energy, Development Research Group. We would also like to thank Helen Mountford, Ronald Steenblik, Robertus Dellink, and Jean Chateau, all of the Organisation for Economic Co-operation and Development.

The report would not have been possible without the invaluable advice, comments, and suggestions of the Expert Advisory, including Daniel Kammen, Chief Technical Specialist for Renewable Energy and Energy Efficiency, Sustainable Energy Department, and John Besant-Jones, Lead Energy Consultant, Energy Unit, Sustainable Development Network, and focal points from the Regional Energy Units, including Ani Balabanyan, Franz Gerner, and Kari Nyman (ECA); Migara Jayawardena and Dejan Ostojic (East Asia and Pacific); Roberto Aiello, Susan Bogach, Ariel Yepes, and Todd Johnson (Latin America and the Caribbean); Husam Beides, Silvia Pariente-David, and Vlado Vucetic (Middle East and North Africa); Rashid Aziz, Kwawu Gaba, Ashish Khanna, and Sheoli Pargal (SAR); and Mudassar Imran, Sunil Mathrani, and Erik Fernstrom (Africa).

Abbreviations

ANRE	National Energy Regulatory Agency (Moldova)
BISP	Benazir Income Support Programme (Pakistan)
CAMMESA	Compañía Administradora del Mercado Mayorista Eléctrico (Argentina)
CCT	conditional cash transfer
CDC	Caisse de Compensation (Morocco)
CFE	Comisión Federal de Electricidad (Federal Electricity Commission; Mexico)
CNE	Comisión Nacional de Energía (Chile)
ECA	Excess Crude Account (Nigeria)
EDE	electricity distribution company (Argentina)
EML	Electricity Market Law (Turkey)
EMRA	Energy Market Regulatory Authority (Turkey)
EU	European Union
FY	fiscal year
GDP	gross domestic product
GPOBA	Global Partnership on Output-Based Aid
IEPS	Impuesto Especial de Productos y Servicios (special tax on products and services; Mexico)
IMF	International Monetary Fund

kWh	kilowatt-hour
LEAP	Livelihood Empowerment against Poverty (Ghana)
LNG	liquefied natural gas
LPG	liquefied petroleum gas
MDG	Millennium Development Goal
MEM	Mercado Eléctrico Mayorista (Wholesale Electricity Market; Argentina)
MW	megawatts
NAF	National Aid Fund (Jordan)
NAPEP	National Poverty Eradication Program (Nigeria)
NPA	National Petroleum Authority (Ghana)
OECD	Organisation for Economic Co-operation and Development
OMG	oil marketing company
ONE	Office National de l'Electricité (Morocco)
PEMEX	Petróleos Mexicanos
PFB	Poverty Family Benefit Program (Armenia)
PLN	Perusahaan Listrik Negara (Indonesia)
PRA	Programa de Reducción de Apagones (Dominican Republic)
PURC	Public Utilities Regulatory Commission (Ghana)
PWP	Public Works Project (Republic of Yemen)
SFD	Social Fund for Development (Republic of Yemen)
SOCAR	State Oil Company of the Azerbaijan Republic
SRMP	Social Risk Mitigation Program (Turkey)
SWF	Social Welfare Fund (Republic of Yemen)
TEDAS	Turkish Electricity Distribution Co.
TNB	Tenaga Nasional (Malaysia)
TOR	Tema Oil Refinery (Ghana)
TSA	targeted social assistance (Azerbaijan)
UCT	unconditional cash transfer

Overview

Introduction

Poorly implemented energy subsidies are economically costly to taxpayers and damage the environment through increased emissions of greenhouse gases and other air pollutants. Energy subsidies also create distortive price signals and result in higher energy consumption or production as well as barriers to entry for cleaner energy services. Subsidies to consumption, by lowering end-use prices, can encourage increased energy use and reduce incentives to conserve energy efficiently.

Universal energy-price subsidies tend to be regressive because benefits are conditional upon the purchase of subsidized goods and increase with expenditure. The proportional adverse impact of energy subsidy removal can be greatest for the poor, even though the rich receive most of the total value of the subsidy (IEA, OPEC, OECD, and World Bank 2010).

Since the declaration of intent at the G-20 Summit held in Pittsburgh in September 2009, considerable global momentum has been building to phase out fossil fuel subsidies, with many countries already implementing or announcing plans to do so. This evolution is part of an effort to make markets more efficient by matching tariffs more accurately to reflect full costs. The G-20 Communiqué of the Meeting of Finance Ministers and Central Bank Governors, which was held in Gyeongju, Korea, on October

23, 2010, "note[d] the progress made on rationalizing and phasing out inefficient fossil fuel subsidies and promoting energy market transparency and stability and agreed to monitor and assess progress towards this commitment at the Seoul Summit" (G-20 2010).

This report aims to provide the emerging lessons from a representative sample of case studies that could help policy makers to address implementation challenges, including overcoming political economy and affordability constraints, by looking at complementary instruments to compensate vulnerable groups for energy-price increases.

Sample Selection

This report selected a representative sample of case studies in 20 developing countries, based on a number of criteria, including the countries' level of development (and consumption) and energy dependency (distinguishing between net energy exporters and importers), as displayed in table O.1.

The case studies have been selected on the hypothesis that energy dependence and per capita income appear to be the key drivers of subsidy reforms in developing countries. Of the two criteria, energy dependence is expected to be the most powerful determinant of the choice to engage in energy reforms, whereas the level of per capita income may

Table O.1   Countries Selected for Case Study Analysis of Energy Subsidy Reforms

	Region	Energy net importer	Energy net exporter
		Group A	Group C
Low- and Lower- Middle Income Countries	Africa	Ghana[a]	Nigeria
	E. Asia & Pacific		Indonesia[a]
	Eur. & Cent. Asia	Armenia,[a] Moldova[a]	Azerbaijan
	Mid. East & N. Africa	Morocco,[a] Jordan[a]	Egypt, Arab Rep.;[a] Iran, Islamic Rep.; Yemen, Rep.[a]
	S. Asia	India,[a] Pakistan[a]	
		Group B	Group D
Upper-Middle and High- Income Countries	E. Asia & Pacific		Malaysia[a]
	Eur. & Cent. Asia	Turkey[a]	
	L. Amer. & the Caribbean	Chile, Dominican Republic, Peru	Argentina,[a] Mexico[a]

Note: Selected sample based on income, region, and energy net import-export.
a. Country is characterized by macro unbalances (either budget deficit higher than 4 percent of gross domestic product [GDP] or public debt higher than 40 percent of GDP).

pose different challenges in relation to the distributional impact of such reforms on the poor. Energy net importers are expected to have more incentives to undertake energy subsidy reforms when the fiscal burden of such subsidies reaches a significant percentage of gross domestic product (GDP), particularly when there are already macro unbalances related to high thresholds of public budget and debt. Low- and middle-income countries are expected to display a larger impact of energy subsidy reforms on consumption. This impact reflects the opportunities to influence future behavior rather than current consumption trends because of inertia, vested interests, and the presence of affordability issues.

Sample Categorization

The performance of countries that have embraced substantial subsidy reform is compared with the counterfactual of countries that have done little to phase out energy subsidies. This comparison is undertaken within each group of countries determined by the taxonomy of energy dependence and country annual per capita income. The four groups of countries are referred to as Groups A, B, C, and D:

- Groups A and B consist of energy net *importer* countries with lower to middle income and upper-middle to high income, respectively.
- Groups C and D represent energy net *exporters* with lower to middle income and upper-middle to high income, respectively.

The case studies were supported by data collection related to direct budgetary subsidies, fuel and electricity tariffs, and household survey data from official documents, complemented by information publicly available through the websites of the countries' finance and energy ministries and energy service providers.

Sample Distribution

The selected sample is balanced in terms of distribution by the two key dimensions: country level of development and energy dependence (see figures O.1 and O.2). The selected countries also cover all six developing-country regions (see figure O.3) and thereby reflect broad regional features such as Africa's low access rates; Europe and Central Asia's full access rates; Latin America and the Caribbean's leadership in market reform; South Asia's high growth in power demand; and the Middle East and North Africa's crucial role as an energy trade crossroad.

Figure O.1   Distribution of Sample Countries by Energy Net Exports or Imports

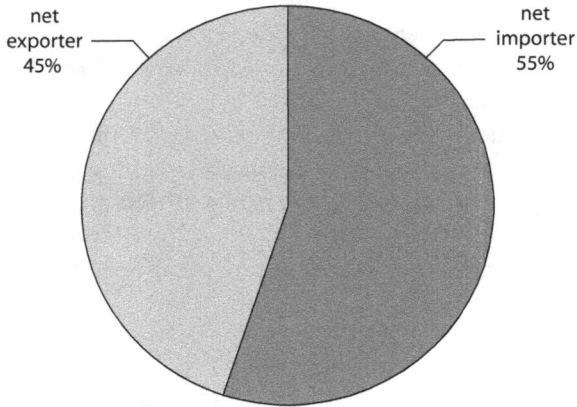

net
exporter
45%

net
importer
55%

Figure O.2   Distribution of Sample Countries by Income

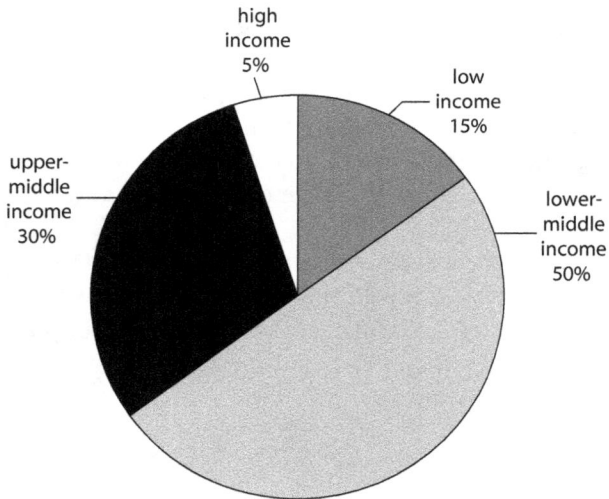

high
income
5%

low
income
15%

upper-
middle
income
30%

lower-
middle
income
50%

Last, as figure O.4 illustrates, the sample also represents a balanced sample in terms of the use of each subsidized fuel: (a) petroleum fuels, including gasoline, diesel, and kerosene; (b) liquefied petroleum gas (LPG); (c) electricity generated from fossil fuels; and (d) natural gas and liquefied natural gas (LNG). Distinction by fuel is of crucial importance,

Figure O.3 Distribution of Sample Countries by Region

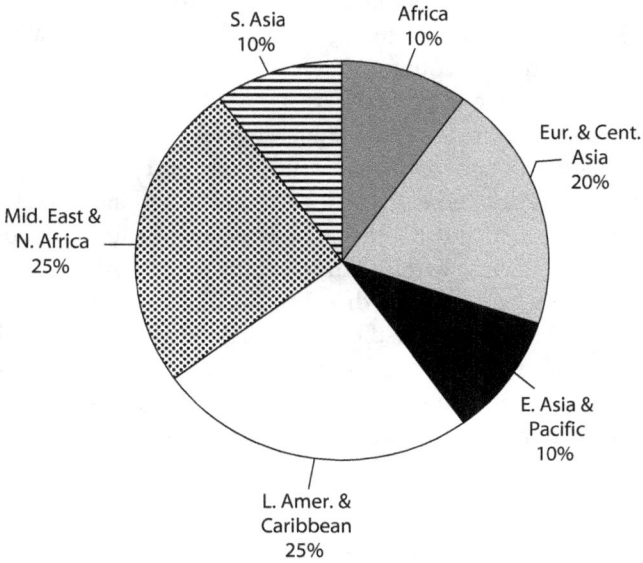

Figure O.4 Distribution of Sample Countries by Fuel Used

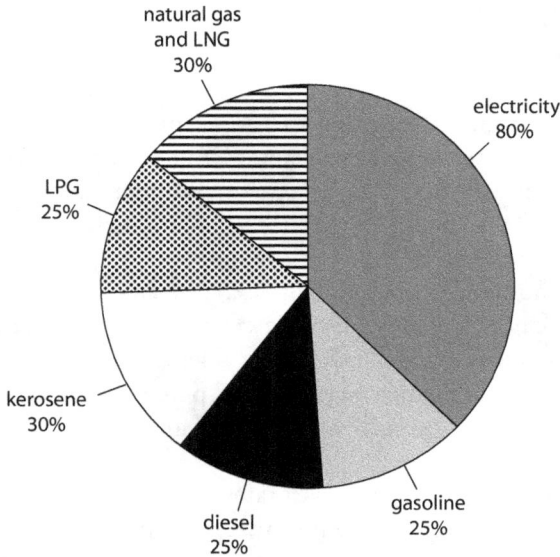

Note: Distribution by fuel used does not sum up to 100 percent because each country can subsidize more than one fuel. "Electricity" reflects only electric power generated from fossil fuels. LPG = liquefied petroleum gas. LNG = liquefied natural gas.

both to reflect the different patterns of consumption by household and to show the potential indirect effect of subsidy reforms.

Typically, kerosene is used for lighting and heating, especially in countries (or within areas of countries) where households do not have access to electricity. Gasoline and diesel are typically used for transport, with gasoline used in internal combustion engines such as motor vehicles (excluding aircraft) and diesel used in goods and passenger transport, agriculture (in pumps and engines, for example), and industry. Diesel and kerosene are close substitutes because large quantities of kerosene can be added to diesel fuel without much impact on vehicle performance. Low kerosene prices relative to diesel thus usually result in the diversion of kerosene to the automotive diesel sector. Adulteration of gasoline with kerosene in other than small quantities can cause damage to vehicles. In the long run, gasoline and diesel are also close substitutes—for example, through the switching from gasoline- to diesel-powered vehicles with an associated worsening of air pollution.

Country Taxonomy, by Macroeconomic and Sectoral Challenges

All groups enhanced the level of GDP per capita, as figure O.5 shows. Countries belonging to different groups experienced different macroeconomic challenges:

- Although Group C recorded the most impressive progress, with a threefold increase in GDP per capita, Group D was characterized by a less significant increase. The plunge of GDP per capita that Group D took in 2002 was mainly due to the economic crisis in Argentina that developed in the early 2000s.
- Although not represented in figure O.5, the annualized growth rate of GDP per capita for lower-income countries (Groups A and C) was about 13 percent, significantly higher than that of the richer countries (Groups B and D), which was around 7 percent. This boost for Group A and C countries reflects a significant improvement from a lower initial situation.
- When comparing average growth rates of net exporters and importers in terms of GDP, it can be observed that net exporters of energy (Groups C and D) grew at a faster rate (but not significantly so) than the net importers (with a 12 percent annualized rate of growth compared with 10 percent, respectively).

Figure O.5   GDP of Sample Countries, by Group

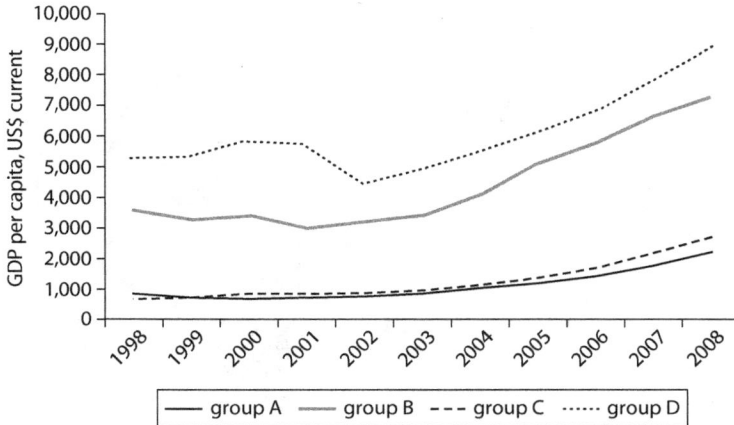

Source: World Bank, World Development Indicators.
Note: Group A countries are energy net importers of lower to middle income. Group B countries are energy net importers of upper-middle to high income. Group C countries are energy net exporters of lower to middle income. Group D countries are energy net exporters of upper-middle to high income.

Differences in Income Equality

In contrast to the positive development in GDP growth, the social challenges coming from income inequality remain considerable. Only Group B recorded a notable decrease in income inequality, with the Gini index dropping from 51 to 47 (see figure. O.6).[1] Groups A and C kept the Gini index almost unchanged. Group D recorded an increase in income inequality, with the Gini index increasing by several points.

The average annualized rate of decline of the Gini index for lower-income countries (Groups A and C) was about 0.4 percent. This is a significantly higher value than that of the richer countries (Groups B and D), for which the average annualized rate of decline was only 0.03 percent. This reflects a significant reduction in inequality in poorer countries. For energy net exporters (Groups C and D), inequality grew by an annual average of 0.1 percent, whereas income inequalities declined for net importers (Groups A and B) by 0.5 percent annually.

Differences in Fiscal Balance

Figure O.7 illustrates the fiscal challenges of the different groups, all subject to significant fluctuations over time, with the exception of Group A. Despite the highly volatile pattern over time, lower-income net energy

Figure O.6   Gini Index for Sample Countries, 1998–2008

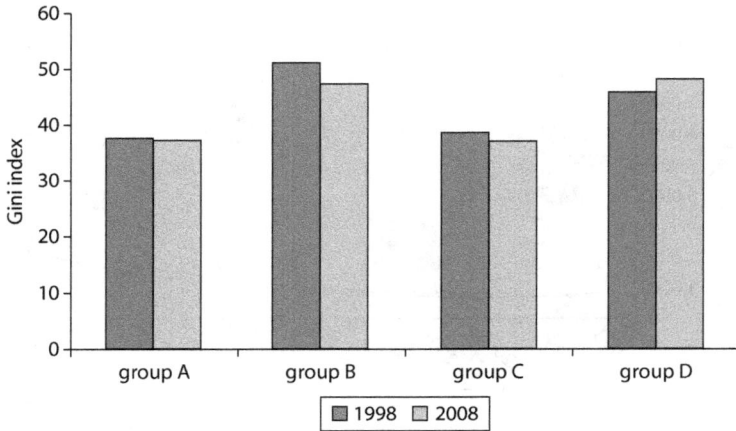

Source: World Bank, World Development Indicators.
Note: Group A countries are energy net importers of lower to middle income. Group B countries are energy net importers of upper-middle to high income. Group C countries are energy net exporters of lower to middle income. Group D countries are energy net exporters of upper-middle to high income. The Gini index measures inequality of income or consumption, here varying between 0 for complete equality and 100 for complete inequality.

Figure O.7   General Government Net Lending, Sample Countries, 1998–2008

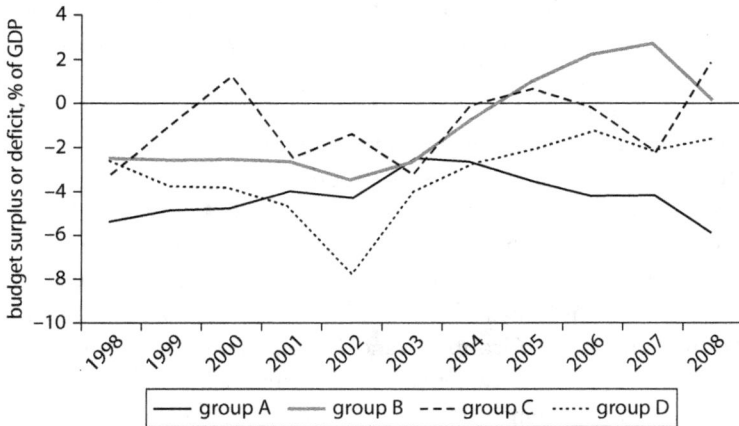

Source: World Bank, World Development Indicators.
Note: Group A countries are energy net importers of lower to middle income. Group B countries are energy net importers of upper-middle to high income. Group C countries are energy net exporters of lower to middle income. Group D countries are energy net exporters of upper-middle to high income.

exporter (Group C) countries—on average—improved their fiscal situation from a budgetary deficit equal to 3.2 percent of GDP in 1998 to a budget surplus of 1.8 percent of GDP in 2008. Higher-income net importers (Group B) also moved from a deficit to a surplus. Group B recorded the most notable improvement in its fiscal situation. Upper-middle to high-income net energy exporters (Group D) displayed the sharpest deterioration in the budgetary situation, brought about by the Latin American crisis, but their fiscal status has improved since then.

Between 1998 and 2008, there was a clear declining trend in government gross debt for all groups of countries except Group D, as figure O.8 shows. In fact, Group D countries increased their public debt on average from 40 percent of GDP in 1998 to about 50 percent in 2008. In the aftermath of the Argentinean economic crisis, public debt of Group D even reached up to 85 percent of GDP (2002/03).

At the same time, Group C improved its fiscal situation by reducing its general government gross debt annual rate of almost 7.5 percent, cutting it from about 60 percent in 1998 to 30 percent in 2008. Similarly, Group A countries reduced their public debt from 89 percent of GDP to 51 percent. This progress underlines the efforts and headway that

Figure O.8   General Government Gross Debt, Sample Countries, 1998–2008

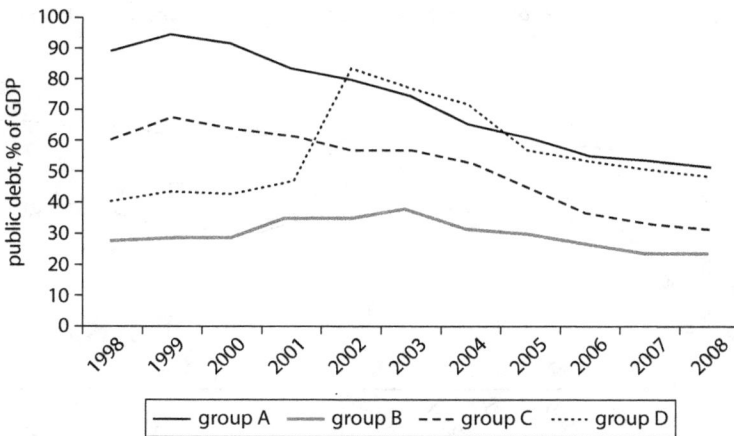

Source: World Bank, World Development Indicators.
Note: Group A countries are energy net importers of lower to middle income. Group B countries are energy net importers of upper-middle to high income. Group C countries are energy net exporters of lower to middle income. Group D countries are energy net exporters of upper-middle to high income.

lower-income countries (Groups A and C) made. When comparing importing (Groups A and B) and exporting (Groups C and D) countries, no notable difference in their fiscal performances can be detected. Both reduced public debt by an annualized rate of 4 percent.

Differences in Energy Production and Use

All countries increased the share of electricity generated using fossil fuels as primary energy, as figure O.9 shows. Between 1998 and 2008, Group C and D countries (energy net exporters) increased such a share from 83 percent to 90 percent and from 74 percent to 80 percent, respectively. For A and B countries (energy net importers), the overall increase was less pronounced. Group A kept the share of electricity generated from fossil fuels unchanged at 70 percent, while Group B countries increased it from 56 percent to 60 percent. Also, on average, higher-income countries (Groups B and D) increased the production of electricity from fossil fuels only at a slightly higher rate than lower-income countries (Groups A and C).

Energy exports have decreased for Groups C and D, and imports remained constant for Groups A and B, as figure O.10 shows. Group D displays the sharpest decline in energy exports (reflected in figure O.10

Figure O.9   Electricity Production from Fossil Fuels, Sample Countries, 1998–2008

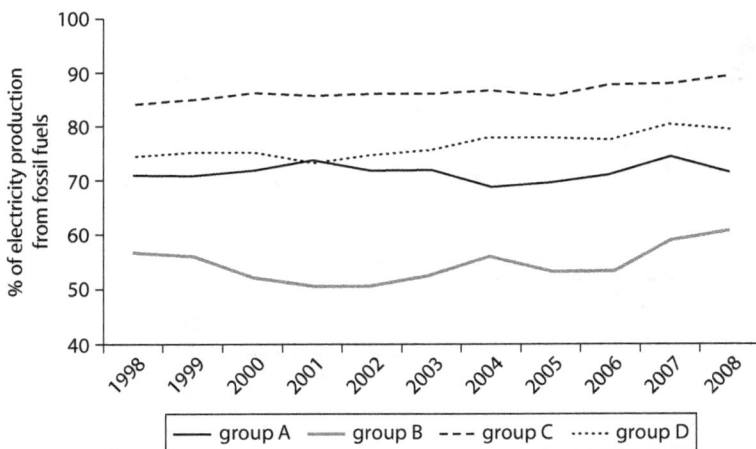

Source: World Bank, World Development Indicators.
Note: Group A countries are energy net importers of lower to middle income. Group B countries are energy net importers of upper-middle to high income. Group C countries are energy net exporters of lower to middle income. Group D countries are energy net exporters of upper-middle to high income.

Figure O.10   Energy Net Imports, Sample Countries, 1998–2008

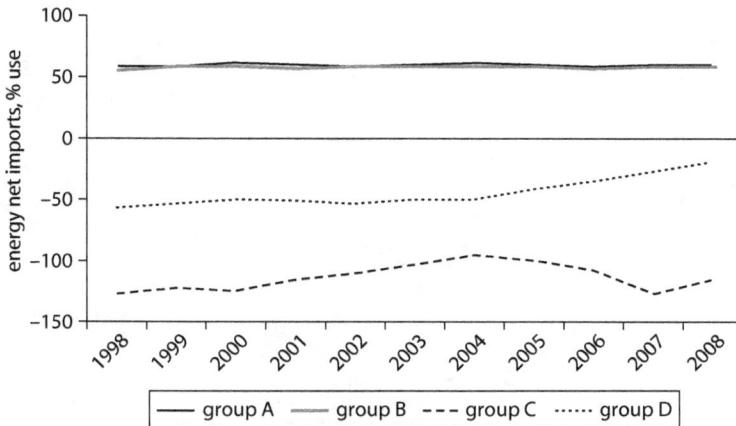

Source: World Bank, World Development Indicators.
Note: Group A countries are energy net importers of lower to middle income. Group B countries are energy net importers of upper-middle to high income. Group C countries are energy net exporters of lower to middle income. Group D countries are energy net exporters of upper-middle to high income.

as negative imports), from 50 percent in 1998 to 4 percent in 2008, which represents an annualized rate of change of almost 10 percent. When comparing higher-income with lower-income countries, one can see that the higher-income countries (Groups B and D) tended to increase the net imports at a faster rate than the lower-income countries (Groups A and C).

Country Taxonomy, by Success in Energy Subsidy Reform

Some interesting patterns emerge by grouping countries in our taxonomy, distinguishing countries according to their energy dependence and the level of income. As one would expect, net energy importers (Groups A and B) have reduced more significantly the burden of energy subsidies in the budget (see figure O.11). On the other hand, for net energy exporters (Groups C and D), the burden of energy subsidy in the budget increased only slightly overall, although it doubled for the higher-income countries (Group D).

Considering the trends in gasoline and diesel retail prices, net energy importers (Groups A and B) have also been the most successful in phasing out subsidies by increasing the price of petroleum product prices, as figures O.12 and O.13 show.

Figure O.11   Budgetary Energy Subsidy in Sample Countries, 2004–10

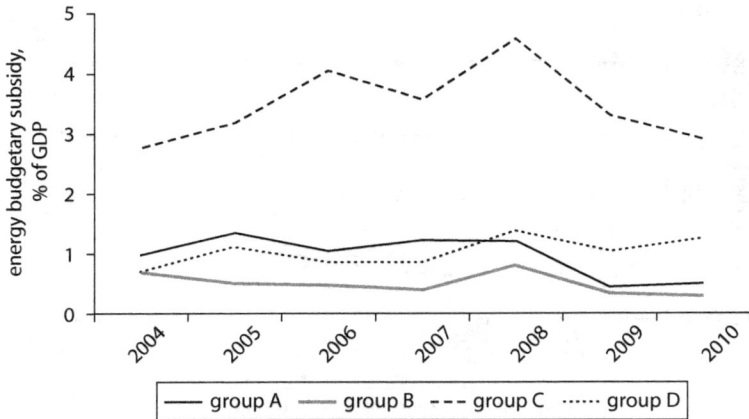

Sources: Based on IMF (various years) and available information from sample countries' ministries of finance.

Figure O.12   Gasoline Retail Tariffs in Sample Countries, 2002–10

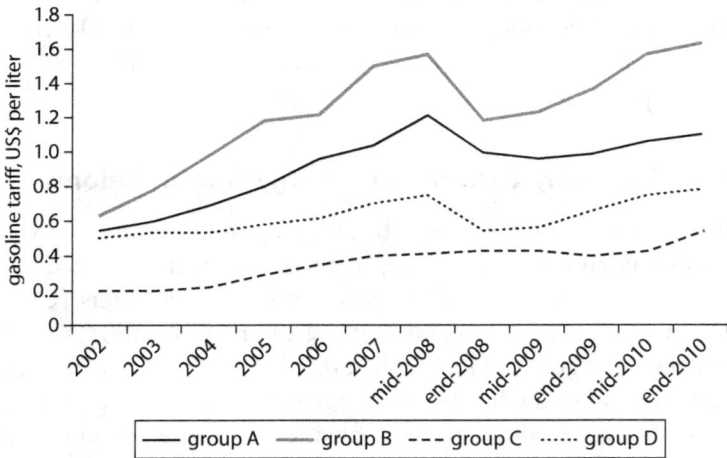

Sources: Based on data from GIZ n.d.; IMF 2010; and data from individual sample countries' information.

Figure O.13   Diesel Retail Tariffs in Sample Countries, 2002–10

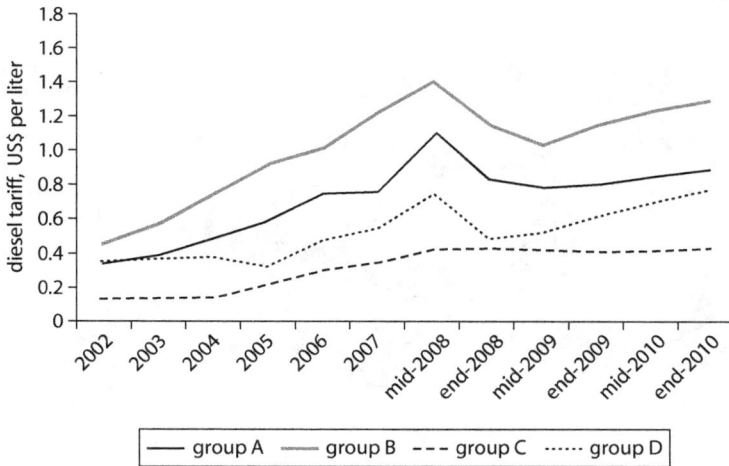

Sources: Based on data from GIZ n.d.; IMF 2010; and data from individual sample countries' information.

Targeting Subsidy Performance

To measure the performance of a subsidy in reaching the poor, policy makers may find it helpful to define the probability that the targeted group (in this case, the poor) will receive the subsidy. This index is known as the beneficiary incidence. Among the overall poor population, policy makers may find that a decomposition of the beneficiary incidence enables some quick diagnostics of the key problems and the required policy responses to be derived. Such a decomposition, listed below using the example of electricity, breaks down into three components (as figure O.14 illustrates):

- *The share of households with potential access to the energy source (A).* This is determined by the coverage of the electricity grid among the population, which is in turn influenced by the development of the infrastructure grid network and its geographical reach (within a reachable distance from where households live). If A is low, which is often the case in rural areas, the best policy response is to develop the most suitable infrastructure (including off-grid and rural electrification solutions) to reach the poor.
- *The share of households with access that actually use the energy source (U).* This component captures both supply- and demand-side variables.

Figure O.14   Beneficiary Incidence: How Much of the Poor Does the Subsidy Reach?

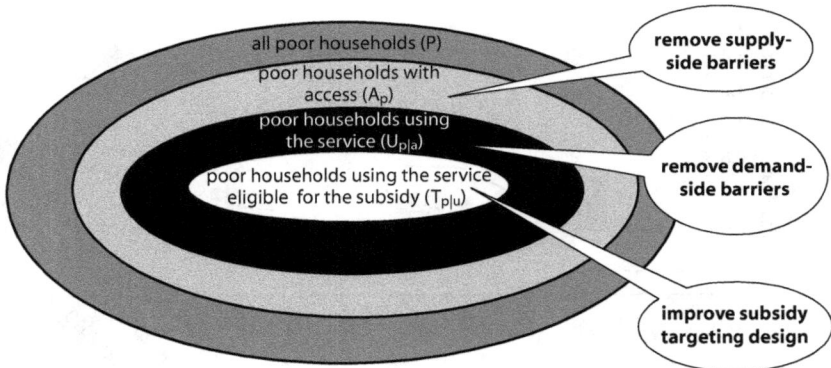

all poor households (P)

poor households with access (A_p)

poor households using the service ($U_{p|a}$)

poor households using the service eligible for the subsidy ($T_{p|u}$)

remove supply-side barriers

remove demand-side barriers

improve subsidy targeting design

Source: Adapted from Wodon 2009.

A low value of U can result from affordability constraints due to the expensive connection rates; from the presence of cheaper but dirtier substitutes (which can in turn cause environmental problems); or both.

- *The share of households that are connected to the energy source and are eligible for the subsidy (T).* The third factor is determined by eligibility criteria included in the subsidy design. More sophisticated schemes, such as the U.S. Low Income Home Energy Assistance Program (LIHEAP), base eligibility criteria on socioeconomic variables.

Beyond the beneficiary incidence defined above, additional components to be considered are the following (see figure O.15):

- *The rate of subsidization*—calculated from the ratio between household consumption (valued at cost-recovery prices) and the actual payment— among those who benefit from the subsidy. This component can be improved by better targeting of the subsidy design.
- *The quantity consumed* among those who benefit from the subsidy, which depends mainly on income. Jacobson, Milman, and Kammen (2005) show how electricity consumption is far more evenly distributed in developed countries than in developing countries, suggesting that the distributional pattern of electricity consumption depends heavily on a combination of wealth, income distribution, and quality of

Figure O.15   Benefit Incidence of Subsidies

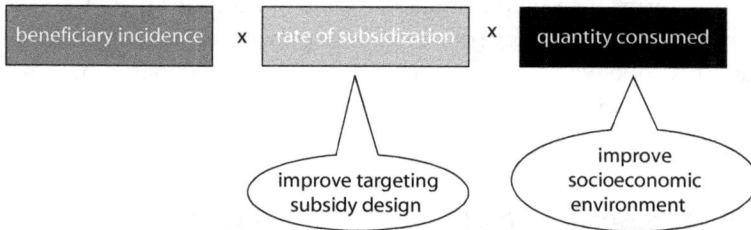

Source: Adapted from Wodon 2009.

infrastructure provision. Metrics relating energy access to income pro-vide a quantitative basis to evaluate the effectiveness of pricing reforms in meeting economic efficiency, social equity, and environmental goals (Jacobson, Milman, and Kammen 2005).

Lessons from the Case Studies

The review of country experiences highlights some lessons that may help other countries to engage in such reforms, particularly when belonging to lower income groups.

Strengthen Social Safety Nets and Improve the Targeting Mechanisms for Subsidies

Use of transitional arrangements and short-term measures to alleviate the impact of tariff increases on the poor can act to protect low-income groups at the time of the policy change. Policy tools to protect the poor include lifeline rates, which generally perform better than universal sub-sidies, as well as cash transfers. Targeting mechanisms and methods for identifying those eligible for the subsidy program can vary, depending on the degree of coverage as well as the extent to which different programs benefit the poor, determining trade-offs between different solutions. To be effective, subsidy programs should adopt simple and transparent targeting criteria consistent with those adopted by the social assistance system.

Inform the Public and Announce One-Off Compensatory Measures

Governments need to ensure public trust in the reform agenda through broad communication, appropriate timing of subsidy removal, and implementation of compensatory social policies. Planning careful com-munication strategies—including media and public campaigns in order to reach out to the poor and those who will be most affected by the subsidy reform—can help minimize public opposition to energy subsidy reforms. While developing social safety nets is important in ensuring that consumers cope with higher prices successfully in the long run, tariff and fuel price increases should be accompanied by immediate short-term measures.

Ensure the Sustainability of Subsidy Policy through Broader Sectoral Reforms

In the power sector, engaging in broader reforms to improve service ahead of reforming energy subsidies, particularly where the quality of electricity services is low, lends credibility and improves consumer willingness to pay. Improving energy efficiency will also help to reduce the costs of removing subsidies both for energy suppliers and consumers. More generally, rationalizing the fuel mix for electricity and transport—discouraging private transport in favor of public transport—may help to support reforms as well as the prioritization of structural expenses that benefit the poor (including sectoral road and rural electrification schemes but also social expenditure, including health and education).

Structure of the Report

Group-Specific Introductions

The volume is divided into four parts, each pertaining to one of the four country groups (A–D) and comprising the country case studies within each group. Opening each part is an introductory discussion of the group's key macroeconomic and social challenges and the dependence of each economy on fossil fuels.

- *Income and inequality trends,* captured by the selected indicators:
  - GDP (US$ current per capita) divided by mid-year population[2]
  - Gini index (1–100) of economic inequality.[3]
- *Fiscal challenges,* captured by the following indicators:
  - General government net lending or borrowing[4]
  - General government gross debt (% GDP).[5]
- *Fossil fuel dependency,* illustrated by these indicators:
  - Electricity production from fossil fuels[6]
  - Net energy imports.[7]

Country Case Studies

Each of the case studies is divided into four key sections, to facilitate comparison:

- *"Incentives to Energy Subsidy Reforms"* describes the key incentives and drivers for energy subsidy reforms, drawing from the evidence collected on the fiscal burden of energy subsidies, both directly in the budget and implicitly, using evidence from the hidden costs of subsidies or any other relevant evidence, as from these indicators:
  - Energy explicit budgetary subsidy (% GDP)[8]
  - Energy implicit subsidies (% GDP).[9]
- *"Reform Efforts"* describes the country's energy subsidy reforms based on the evidence on removing fossil fuel subsidies by increasing petroleum and electricity prices and other nonprice reforms. The section also covers the impact on road fuel consumption and power consumption, looking at trends in these indicators:
  - Fossil fuel prices (US$ per liter)[10]
  - Road fuel consumption (kt of oil equivalent per capita)[11]
  - Electricity prices (US$ per kilowatt-hour [kWh])[12]
  - Power consumption (kWh per capita).[13]
- *"Poverty Alleviation Measures"* describes in detail the evidence from household surveys on the distributional impact of subsidies by income quintile (or decile, if available) as well as the materiality of energy subsidies—reporting for each fuel how much income is spent by each quintile of the population, using the following indicators (which, except for block tariff data, come from household budget and expenditure surveys):
  - Electricity block tariffs (US$ per kWh)[14]
  - Access to different energy sources (what proportion of households is connected to each energy source), by quintile or decile (where available), divided by rural and urban areas
  - Expenditure on different energy sources (what proportion of income poor households spend on each fuel), by quintile or decile (where available), divided by rural and urban areas
  - Beneficiary incidence of energy subsidies (how well the subsidy targets benefits to poor households as opposed to other households), by quintile or decile (and by fuel type, where available)
  - Beneficiary incidence of energy subsidies (what proportion of poor households as a whole receive the subsidy), by quintile or decile (and by fuel type, where available)

    ◦ Welfare impact of removing energy subsidies, by fuel, by quintile or decile, or by both.

The section also reports any evidence from social safety nets or alternative mechanisms that were used in the implementation of energy subsidy reforms to protect the poor.

- *"Key Lessons Learned"* sums up the conclusions from the case study and describes the lessons a particular country may offer to others that seek to reform energy subsidies.

Notes

1. The Gini-coefficient—the most commonly used measure of inequality of income or consumption—varies between 0, which reflects complete equality, and 1 (equivalent to 100 in figure O.6), which indicates complete inequality (World Bank, *World Development Indicators*).

2. GDP is the sum of gross value added by all resident producers in the economy plus any product taxes and minus any subsidies not included in the value of the products. It is calculated without making deductions for depreciation of fabricated assets or for depletion and degradation of natural resources (World Bank, *World Development Indicators*).

3. The Gini index measures the extent to which the distribution of income (or consumption expenditure) among individuals or households within an economy deviates from a perfectly equal distribution. The Gini index measures the area between the Lorenz curve (which plots the cumulative percentages of total income received against the cumulative number of recipients, starting with the poorest individual) and a hypothetical line of absolute equality, expressed as a percentage of the maximum area under the line. Thus a Gini index of 0 represents perfect equality, while an index of 100 implies perfect inequality (World Bank, *World Development Indicators*).

4. Net lending (+) or borrowing (−) is calculated as revenue minus total expenditure. This is a core Government Finance Statistics balance that measures the extent to which general government is either putting financial resources at the disposal of other sectors in the economy and nonresidents (net lending) or using the financial resources generated by other sectors and nonresidents (net borrowing). This balance may be viewed as an indicator of the financial impact of general government activity on the rest of the economy and nonresidents (IMF 2001, paragraph 4.17). Note: Net lending (+) or borrowing (−) is also equal to net acquisition of financial assets minus net incurrence of liabilities.

5. Gross debt consists of all liabilities that require payment or payments of interest or principal by the debtor to the creditor at a date or dates in the future. This includes debt liabilities in the form of Special Drawing Rights, currency and deposits, debt securities, loans, insurance, pensions and standardized guarantee schemes, and other accounts payable. Thus, all liabilities in the 2001 Government Finance Statistics Manual system are debt except for equity and investment fund shares and financial derivatives and employee stock options. Debt can be valued at current market, nominal, or face values (IMF 2001, paragraph 7.110).

6. Electricity production from fossil fuels includes oil, gas, and coal sources (as a percentage of total electricity). Total electricity production is measured at the terminals of all alternator sets in a station. In addition to hydropower, coal, oil, gas, and nuclear power generation, it covers generation by geothermal, solar, wind, and tide and wave energy, as well as that from combustible renewables and waste (IEA 2010).

7. Net energy imports are estimated as energy use less production, both measured in oil equivalents. A negative value indicates that the country is a net exporter. Energy use refers to use of primary energy before transformation to other end-use fuels, which is equal to indigenous production plus imports and stock changes, minus exports and fuels supplied to ships and aircraft engaged in international transport (IEA 2010).

8. Explicit budgetary energy subsidies are those reported as an item of the expenditure in the consolidated general budget (% GDP) (IMF reports and country ministries of finance).

9. Where available, explicit budgetary expenditures are complemented by estimates by the World Bank and IMF of implicit forms of subsidies. Hidden cost is defined as the difference between actual receipts and the revenue that the energy company (for example, a utility involved in the distribution of electric and natural gas) would receive were it to be in operation with cost-recovery tariffs based on efficient operation with normal losses and with full bill collection (Ebinger 2006 for Eastern Europe and Central Asia; AICD Database for Sub-Saharan Africa).

10. Gasoline and diesel retail tariffs refer to the pump prices of the most widely sold grade of diesel gasoline fuel (US$ per liter). Kerosene retail tariffs refer to the most widely sold grade of kerosene (US$ per liter) (GIZ n.d.; IMF reports; and country oil refineries).

11. Road fuel consumption refers to gasoline and diesel fuel consumption (kt of oil equivalent per capita) used in internal combustion engines, such as motor vehicles, excluding aircraft (in the case of gasoline) and diesel engines (in the case of diesel) (IEA 2010).

12. Electricity prices refer to the residential average electricity tariff levels (US$ per kWh) based on data from country energy regulatory agencies, energy ministries, and electricity utilities.

13. Electric power consumption (kWh per capita) measures the production of power plants and combined heat and power plants less (a) transmission, distribution, and transformation losses; and (b) own use by heat and power plants (World Bank, *World Development Indicators*).

14. Electricity block tariffs are residential average electricity tariff levels (US$ per kWh), by block of consumption (defined by thresholds of kWh) based on data from country energy regulatory agencies, energy ministries, and electricity utilities.

References

AICD (Africa Infrastructure Country Diagnostic) (database). n.d. World Bank, Washington, DC. http://www.infrastructureafrica.org/tools/data.

Ebinger, Jane. 2006. "Measuring Financial Performance in Infrastructure: An Application to Europe and Central Asia." Policy Research Working Paper 3992, World Bank, Washington, DC.

G-20 (Group of 20 Finance Ministers and Central Bank Governors). 2010. "Communiqué, Meeting of Finance Ministers and Central Bank Governors." Gyeongju, Republic of Korea. October 23.

GIZ (German Agency for International Cooperation). n.d. International Fuel Prices database. GIZ (formerly GTZ), Bonn. http://www.gtz.de/en/themen/29957.htm.

IEA (International Energy Agency). 2010. *World Energy Outlook 2010*. Paris: Organisation for Economic Co-operation and Development/IEA.

IEA (International Energy Agency), OPEC (Organization of the Petroleum Exporting Countries), OECD (Organisation for Economic Co-operation and Development), and World Bank. 2010. "Analysis of the Scope of Energy Subsidies and Suggestions for the G-20 Initiative." Joint report submitted to the G-20 Summit Meeting, Toronto, June 26-27. http://www.g20.org/images/stories/canalfinan/gexpert/energia/03energysubsidy.pdf.

IMF (International Monetary Fund). 2001. *Government Finance Statistics Manual 2001*. Washington, DC: IMF.

———. 2010. Retail domestic fuel prices data sheet, IMF, Washington, DC. http://www.imf.org/external/pubs/ft/spn/2010/data/spn1005.csv.

———. Various years. "Article IV Consultations." Country reports, IMF, Washington, DC.

Jacobson, A., A. Milman, and D. Kammen. 2005. "Letting the (Energy) Gini Out of the Bottle: Lorenz Curves of Cumulative Electricity Consumption and Gini Coefficients as Metrics of Energy Distribution and Equity." *Energy Policy* 33 (14): 1825–32.

Wodon, Quentin. 2009. "Electricity Tariffs and the Poor: Case Studies from Sub-Saharan Africa." Working Paper 12, Africa Infrastructure Country Diagnostic, World Bank, Washington, DC.

PART 1

Group A Countries:
Net Energy Importer and Low Income

Macroeconomic and Social Challenges

- Whereas all countries are characterized by an increasing level of income, as displayed by a buoyant growth in gross domestic product (GDP) per capita, inequality has increased or remained constant over time, with the notable exception of Armenia and Jordan, the two countries where energy subsidy reforms were most successful.
- Armenia displays one of the highest GDPs per capita and the lowest Gini index in Group A (see figure P1.1). The South Asian countries in this group, India and Pakistan, are at the bottom by the level of GDP per capita but are characterized by some of the lowest degrees of income inequality (as reflected by the Gini index, see figure P1.2).
- The majority of countries are characterized by a decreasing or stable budget and public debt over time. Moldova made substantial progress in reducing both the budget deficit and public debt. Morocco stands out as the only country displaying a budget surplus and with a substantial reduction in public debt. Jordan substantially reduced both the budget deficit and public debt. In most of the other countries, the picture is mixed. In Armenia and Ghana, the fiscal situation became more challenging, but public debt is substantially reduced (see figures P1.3 and P1.4).

Fossil Fuel Dependence

- All countries with the exception of Armenia have either increased or kept constant the percentage of electricity generated from fossil fuels, as figure P1.5 shows.
- Jordan, Moldova, and Morocco rely almost entirely on electricity generated from fossil fuel production and on substantial imports to satisfy supply. Armenia and Ghana are the countries least dependent on fossil fuel production, even though they both increased the percentage of net imports (see figures P1.5 and P1.6).

Income and Inequality Trends for Group A

Figure P1.1   GDP Per Capita, Group A Countries, 1998–2008

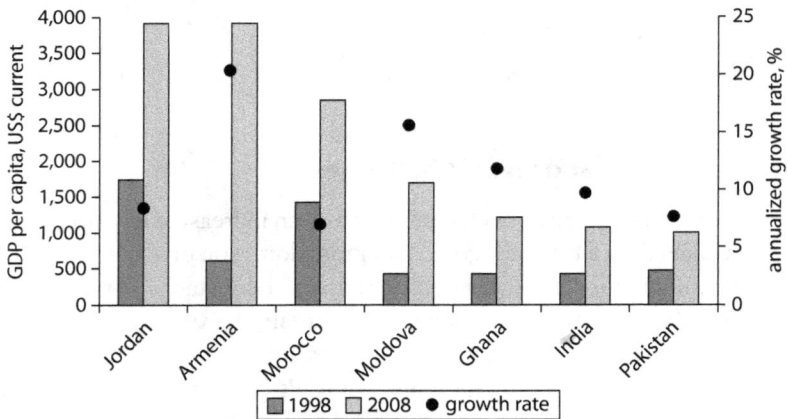

Source: World Bank, World Development Indicators.

Figure P1.2   Gini Index, Group A Countries, 1998–2008

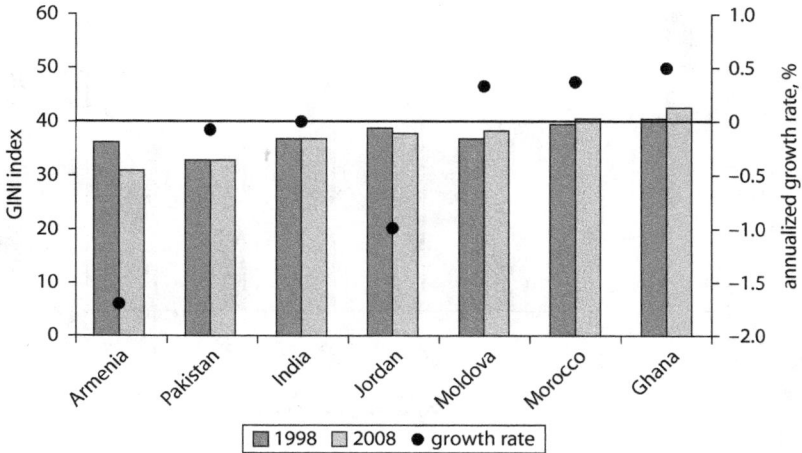

Source: World Bank, World Development Indicators.
Note: The Gini index measures the extent to which the distribution of income (or consumption expenditure) among individuals or households within an economy deviates from a perfectly equal distribution. A Gini index of 0 represents perfect equality, while an index of 100 implies perfect inequality.

Fiscal Indicators for Group A

Figure P1.3   General Government Net Lending or Borrowing, Group A Countries, 1998–2008

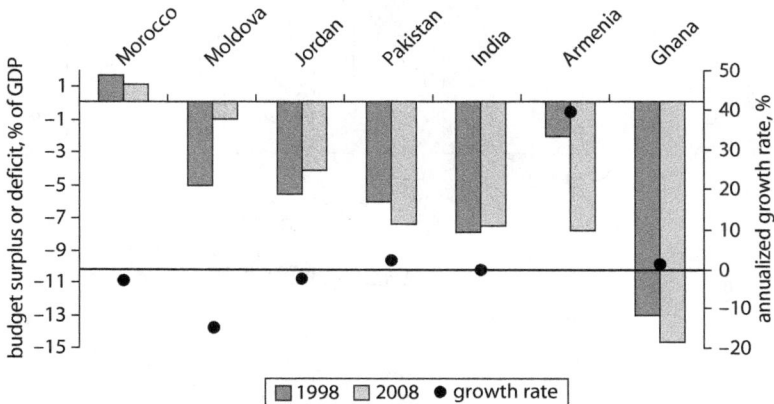

Source: IMF reports, various years.

Figure P1.4   General Government Gross Debt, Group A Countries, 1998–2008

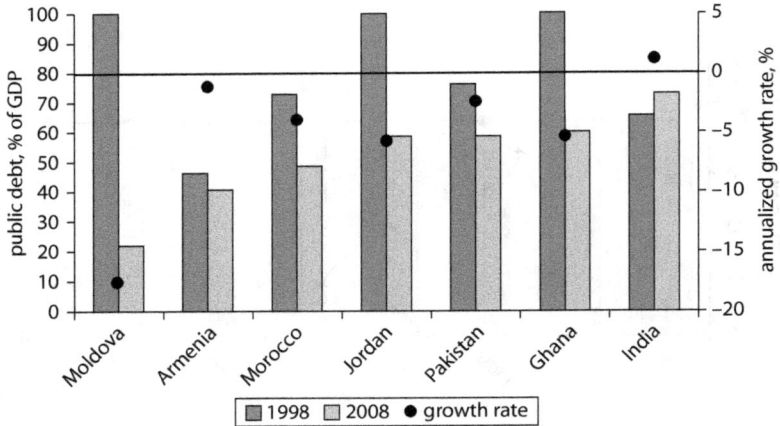

Source: IMF reports, various years.

Fossil Fuel Dependence for Group A

Figure P1.5   Electricity Production from Fossil Fuels, Group A Countries, 1998–2008

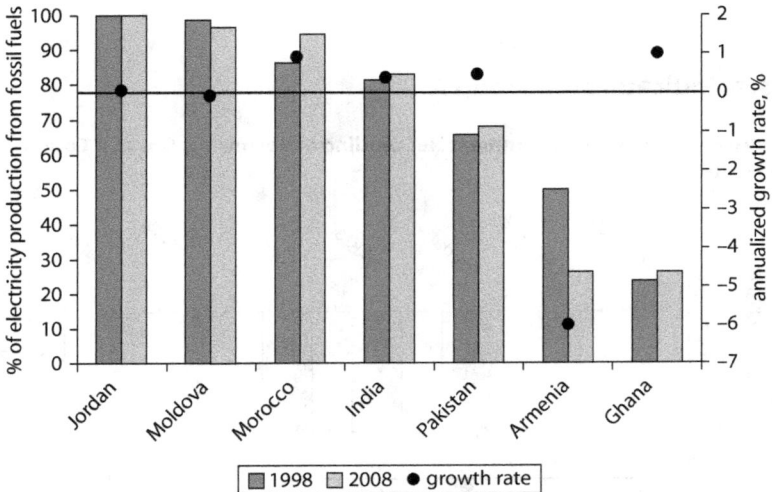

Source: World Bank, World Development Indicators.

Figure P1.6   Energy Net Imports, Group A Countries, 1998–2008

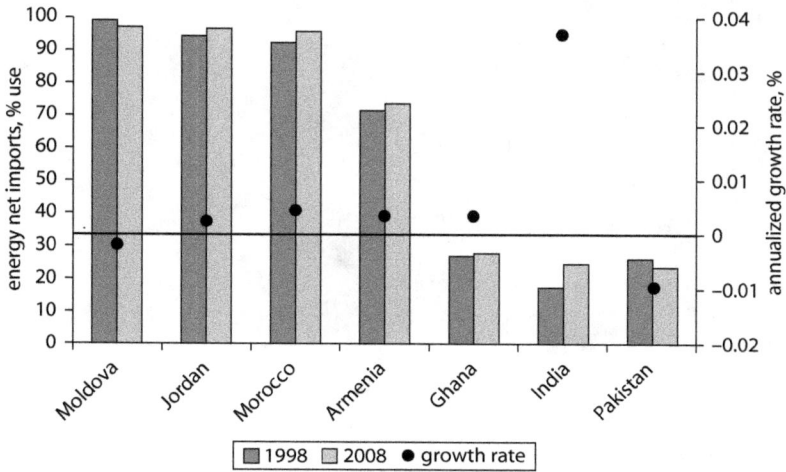

Source: World Bank, World Development Indicators.

CHAPTER 1

Armenia

Incentives to Energy Subsidy Reforms

Armenia's early transition challenges since gaining independence from the former Soviet Union derived from the existence of substantial quasi-fiscal deficits, coming from underpricing. Average electricity tariffs in 1992 and 1993 were roughly one-tenth of the current average electricity tariffs. Fiscal and quasi-fiscal subsidies to the power sector had reached a level equivalent to roughly 11 percent of Armenia's gross domestic product (GDP) in 1995. Collection rates were barely above 50 percent, and the system suffered from severe underinvestment.

This challenging situation provided the government with strong incentives to restructure the power sector. From 1995, the direct budgetary subsidy levels were significantly reduced (as reflected in figure 1A.1 for the 2004–08 period). However, in the early 2000s, even when direct subsidies were very low, consumers still received more than 1 percent of GDP in quasi-fiscal subsidies, as shown in figure 1A.2

In addition to the doubling in the natural gas import prices (from US\$55 per 1,000 cubic meters in 2006 to US\$110 in 2008), the Armenian government decided to use US\$190 million out of the US\$250 million privatization proceeds from the 2008 sale of the Hrazdan-5 power plant to subsidize gas supply. This was widely regarded as a large and poorly targeted subsidization scheme and was abolished in 2008.

Reform Efforts

The power sector reform plans included

- Gradual transition to cost-based tariffs
- Unbundling of part of the state-owned, vertically integrated utility
- Implementation of a new regulatory framework.

Table 1.1 presents a more detailed time line of the power sector reforms.

In some countries, revenues have decreased as a result of government intervention. In Armenia, the government waived the return on assets for state-owned utilities for 2009 and 2010, limiting revenues available for investment. Small hydropower projects have become less attractive because of increased financing costs, and some commercial banks that committed to small hydropower projects funded by international financial institutions are seeking cofinancing sources in Armenian drams (Balabanyan et al. 2011).

Generating companies failed to recover full costs because of the currency devaluation in 2009 and the drop in demand. In addition to the currency devaluation, rising gasoline costs (see figure 1A.3) caused the operating expenditures to increase. Most short-term investments have financing, but medium-term projects still need funding (Balabanyan et al. 2011).

Due to the global economic crisis, energy demand has decreased considerably, but demand in the medium term is still expected to grow by 2–3 percent annually (see figures 1A.4 and 1A.6). This means Armenia will still need to invest in new generation capacity and rehabilitate the existing capacity to continue to meet future needs. Armenia is also looking into diversifying its energy sources not only by expanding its pipeline capacity from Iran, which began functioning in 2008, but also by investing in renewable sources such as geothermal (World Bank 2009; Balabanyan et al. 2011).

Poverty Alleviation Measures

Social Safety Nets

Until January 1999, when the Poverty Family Benefit program (PFB) was put in place, offering cash payments to a targeted group of poor

Table 1.1 Power Sector Reforms in Armenia, 1994–2004

Year	Key events	Collection rate (percent)	Total subsidy (US$, millions)
1994	• Tariff reform begins. Residential and agricultural tariffs are set at 1 Armenian dram per kilowatt-hour (kWh). All other customer tariffs are set at dram 8.80 per kWh, and a schedule is established for gradually raising tariffs for lower-voltage customers.	39	127
1995	• Initial work on the unbundling of the state electricity utility, Armenergo, begins in March with the creation of distinct generation and distribution enterprises. Transmission and dispatch remain within the remit of Armenergo.	54	141
	• Further unbundling begins in December, with the separation of the Hrazdan thermal power plant (TPP) and the Sevan-Hrazdan coordinated hydropower plant (HPP) system from Armenergo.		
1997	• State-owned power sector entities are transformed into 100 percent state-owned closed joint stock companies. Thirteen small HPPs are privatized according to the Law on Privatization.	61	100
	• The Law on Privatization is passed in December, defining the power sector companies and assets to be privatized and corporatized.		
1998	• All 11 electricity distribution companies are consolidated into four large regional distribution companies, established as subsidiaries of the Armenian Electricity Network.	77	66
1999	• Block power tariffs are eliminated in January 1999 in favor of a single end-user tariff of dram 25 per kWh. To compensate for removal of the lowest block (lifeline) tariff, the government reshapes and expands its Poverty Family Benefit program of social transfers to low-income customers.	88	50
	• Payment of electricity bills is shifted to post offices rather than through bill collectors.		
2001	• Implementation of the Automated Metering and Data Acquisition System is completed.	81	63
2002	• In April, the four regional distribution companies are merged into a single company, and all 110-kilovolt substations are transferred to the consolidated Electricity Distribution Company.	90	33
2003	• The Hrazdan TPP and Sevan-Hrazdan HPPs are handed over to the government of the Russian Federation as a repayment of Armenia's state debt to Russia.	96	21
2004	• In February, the Energy Law is further amended to deregulate provision of decentralized heating.	100	0
	• The Law on Energy Efficiency and Alternative Energy is adopted in November to promote the development of renewable energy and to raise energy independence and security in the country.		

Source: Sargsyan, Balabanyan, and Hankinson 2006.
Note: TPP = thermal power plant; HPP = hydropower plant.

households, Armenia had an increasing block tariff structure (see figure 1A.7). The Energy Regulatory Commission eliminated the increasing block tariffs in favor of a single uniform tariff of dram 25 (equivalent to US$0.048 using 1999 conversion rates) per kilowatt-hour (kWh). To soften the impact of the tariff increase, a direct cash transfer of dram 1,450 (approximately US$2.70 using 1999 conversion rates) was provided to approximately 30 percent of households (about 230,000 households) eligible for the family benefit as well as to another 9 percent of households (70,000) expected to have difficulty meeting their electricity payments (Lampietti, Banerjee, and Branczik 2007).

The PFB program is based on proxy means tests. An assessment of its targeting shows that poor households were more than twice as likely to receive the cash transfer as nonpoor households (Lampietti, Banerjee, and Branczik 2007). In addition, households regularly consuming electricity in the first two blocks of the 1998 tariff (0–250 kWh) were significantly more likely to receive the cash transfers. However, only 55 percent of the poor received the cash transfer, meaning that 45 percent of poor households were faced with a 47 percent increase in their electricity tariffs and no mitigating cash transfers. PFB coverage among the bottom consumption deciles improved to 61 percent in 2006 (Lampietti, Banerjee, and Branczik 2007).

The Global Partnership on Output-Based Aid (GPOBA) and the World Bank funded a scheme in Armenia providing grants to poor households for individual heating solutions based on gas heaters and, in some cases, boilers. Affordable and efficient heating is crucial for Armenia because winters are harsh and low-income families allocate up to half of their total annual expenditures to winter heating. The GPOBA funds were disbursed only after the predetermined outputs were met, which provided an incentive for the utility providers to complete the installation in a timely and effective manner.

The project was initiated by the government of Armenia when it requested US$3.1 million from GPOBA for this purpose, which was in addition to the US$3 million provided by the World Bank's Urban Heating Project and the government's own US$530,000. Similar to other cases, GPOBA worked with a local organization that helped administer the scheme and used its established channels to reach out to homeowner associations that had secured commitments from more than half of their residents for the gas heating option (GPOBA 2009).

Natural Gas

More recently, in April 2011, about 60,000 low-income families in Armenia received special coupons entitling them to pay dram 100 per cubic meter instead of the previous dram 132 (about US$0.27 instead of US$0.36). The aim of the governmental initiative is partial compensation of the natural gas tariff consumed in the period of April 1, 2011, to March 31, 2012. The dram 500–600 million (US$1.3–US$1.6 million) planned for the subsidy comes from government compensation funds.

Key Lessons Learned

Privatization brought several benefits through a reduction in the need for implicit and explicit governmental subsidies. The decline of the once-sizable quasi-fiscal subsidies showed direct improvements in the performance of the energy sector. Collection rates increased to close to 100 percent by the end of 2004 while commercial losses decreased to 4 percent of total production. (The impact on the average electricity price can be seen in figure 1A.5.)

Although electricity service quality improved as a result of the 1999 reforms, tariff increases caused some customers to switch to cheaper, often dirtier fuel sources, with costly effects on human health. The World Bank assessed the impact of the 1999 household or residential tariff increase in Armenia on various rural and urban income groups. More than 80 percent of households surveyed said they had substituted other energy sources for electricity, with 60 percent substituting wood fuel and only 24 percent substituting natural gas. A more recent household survey found that 46 percent of the urban population relies on wood and 27 percent on electricity for heating purposes and that many poor households do not heat at all (Sargsyan, Balabanyan, and Hankinson 2006).

Armenia's electricity reform is touted as a success story in part because of the timely execution of the reform and the commendable results that it has delivered. The case of Armenia is especially applicable to countries reforming electricity sectors that were once part of larger and more integrated systems (such as the former Soviet Union's) as well as to countries seeking to curb electricity demand growth and increase efficiency.

Annex 1.1 Armenia Case Study Figures

INCOME LEVEL: Lower-middle income
REGION: Europe and Central Asia
ENERGY NET IMPORTER OR EXPORTER: Net importer
FUEL SUBSIDIES: Electricity, natural gas
PHASING OUT SUBSIDIES: Successful

Fiscal Burden of Energy Subsidy in Armenia

Figure 1A.1   Explicit Budgetary Energy Subsidies in Armenia, 2004–08

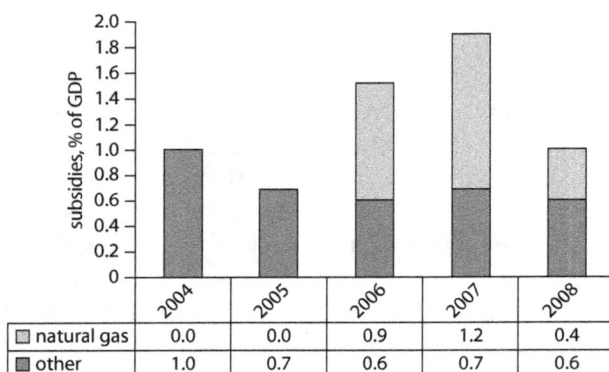

subsidies, % of GDP	2004	2005	2006	2007	2008
☐ natural gas	0.0	0.0	0.9	1.2	0.4
▣ other	1.0	0.7	0.6	0.7	0.6

Source: IMF staff reports, various years.
Note: Budgetary natural gas subsidies equal zero after 2008.

Figure 1A.2   Implicit Subsidies of the Power Sector in Armenia, 2000–03

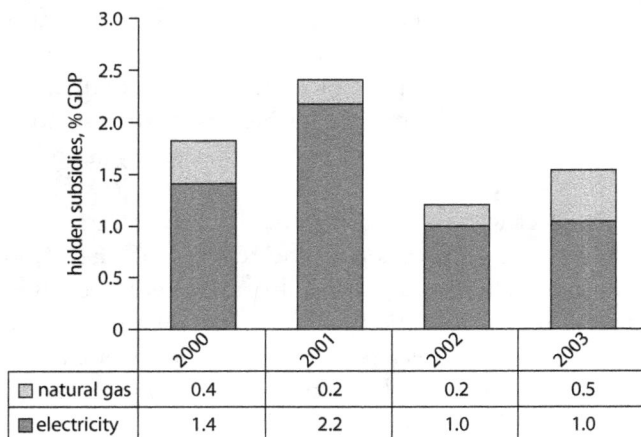

hidden subsidies, % GDP	2000	2001	2002	2003
☐ natural gas	0.4	0.2	0.2	0.5
▣ electricity	1.4	2.2	1.0	1.0

Source: Ebinger 2006.
Note: Implicit subsidies (or hidden costs) are defined as the difference between actual receipts and the revenue that the energy company (for example, a utility involved in the distribution of electric and natural gas) would receive were it to be in operation with cost-recovery tariffs based on efficient operation with normal losses and with full bill collection.

Fuel Prices and Road Sector Consumption in Armenia

Figure 1A.3   Domestic Retail Fuel Prices in Armenia, 2002–10

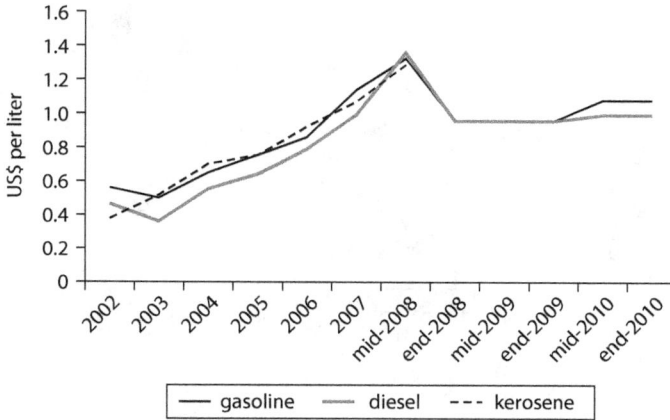

Source: Elaboration of data from GIZ n.d.; IMF 2010; and additional data from individual country information.

Figure 1A.4   Road Sector Diesel Consumption in Armenia, 2000–08

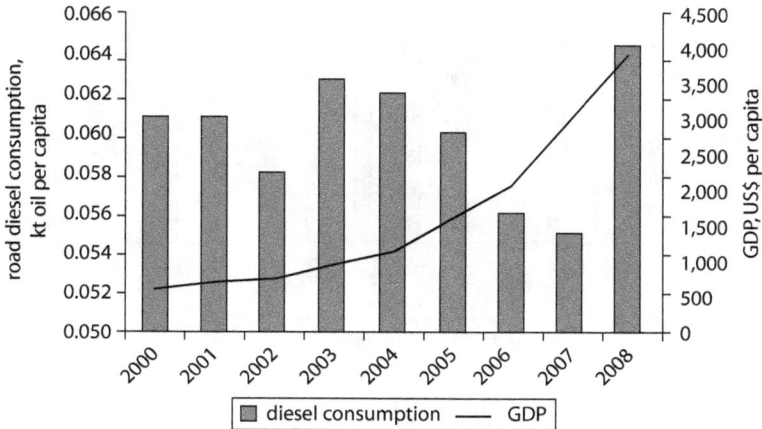

Source: World Bank, World Development Indicators.

Electricity Price and Power Consumption in Armenia

Figure 1A.5   Electricity Price in Armenia, 1998–2010

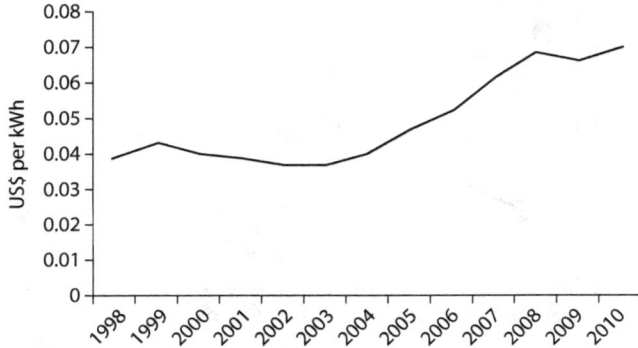

Sources: Elaboration of data from ERRA n.d. and available country information.
Note: kWh = kilowatt-hour.

Figure 1A.6   Power Consumption Per Capita in Armenia, 1998–2008

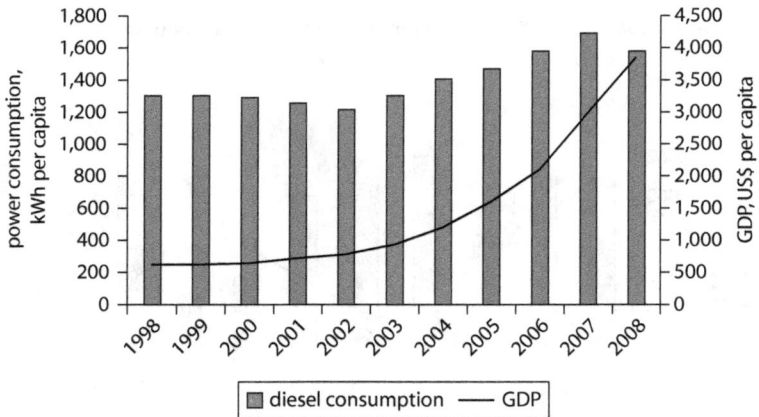

Source: World Bank, World Development Indicators.
Note: kWh = kilowatt-hour.

Poverty Impact Evidence from Household Surveys in Armenia

Figure 1A.7   Electricity Block Tariffs in Armenia, as of 1998

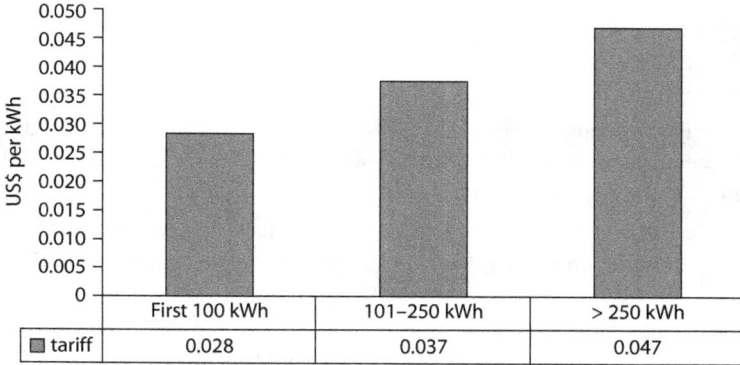

	First 100 kWh	101–250 kWh	> 250 kWh
tariff	0.028	0.037	0.047

Source: Lampietti, Banerjee, and Branczik 2007.
Note: kWh = kilowatt-hour.

Figure 1A.8   Power Consumption in Armenia, by Income Quintile, 2000

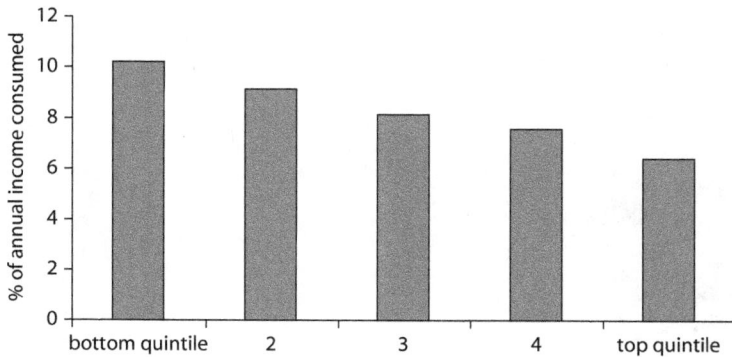

Source: Lampietti, Banerjee, and Branczik 2007.

References

Balabanyan, Ani, Edon Vrenezi, Lauren Pierce, and Denzel Hankinson. 2011. *Outage: Investment Shortfalls in the Power Sector in Eastern Europe and Central Asia.* Directions in Development Series, Washington, DC: World Bank.

Ebinger, Jane. 2006. "Measuring Financial Performance in Infrastructure: An Application to Europe and Central Asia." Policy Research Working Paper 3992, World Bank, Washington, DC.

ERRA (Energy Regulators Regional Association). n.d. Online Electricity & Natural Gas Tariff Database for Central Eastern Europe, Southeast Europe, and the Commonwealth of Independent States, ERRA, Budapest. http://www.erranet.org/Products/TariffDatabase.

GIZ (German Agency for International Cooperation). n.d. International Fuel Prices database. GIZ (formerly GTZ), Bonn. http://www.gtz.de/en/themen/29957.htm.

GPOBA (Global Partnership on Output-Based Aid). 2009. "Output-Based Aid in Armenia: Connecting Poor Urban Houses to Gas Service." Note 23, Europe and Central Asia Region, World Bank, Washington, DC.

IMF (International Monetary Fund). 2010. Retail domestic fuel prices data sheet, IMF, Washington, DC. http://www.imf.org/external/pubs/ft/spn/2010/data/spn1005.csv.

Lampietti, Julian A., Sudeshna Ghosh Banerjee, and Amelia Branczik. 2007. *People and Power: Electricity Sector Reforms and the Poor in Europe and Central Asia*. Directions in Development Series, Washington, DC: World Bank.

Sargsyan, Gevorg, Ani Balabanyan, and Denzel Hankinson. 2006. "From Crisis to Stability in the Armenian Power Sector: Lessons Learned from Armenia's Energy Reform Experience." Working Paper 74, World Bank, Washington, DC.

World Bank. 2009. "Armenia Geothermal Project." Project appraisal document, World Bank, Washington, DC.

CHAPTER 2

Ghana

Incentives to Energy Subsidy Reforms

Despite being classified as an energy net importer, Ghana recently found crude oil off the shores of its western Atlantic coast. At its peak, 20,000 barrels of oil per day could be extracted—which is expected to make Ghana a net oil exporter (World Bank 2009).

The Tema Oil Refinery (TOR) in Ghana produces about 70 percent of Ghana's consumption requirements and uses oil from Nigeria purchased at discounted prices. TOR carries over US$1 billion in debt from refining and distributing petroleum below cost as a result of subsidies. This debt has now been billed as a Debt Recovery Levy to consumers of petroleum products such as gasoline. TOR remains state-owned, which is what some see as the reason why it cannot consistently operate at full cost recovery.

Explicit subsidies to TOR and distributors to compensate for below-cost prices reached 2 percent of gross domestic product (GDP) in 2003 and 2.2 percent in 2004 (Coady et al. 2006). To facilitate the deregulation process, a process for publishing and applying an automatic adjustment formula for pricing petroleum products was completed in 2001 and went into effect in 2003. But it was not until 2005 that the government established the National Petroleum Authority (NPA) to distance itself from petroleum pricing and allow the NPA to monitor and implement the pricing mechanism. Although energy subsidies were reduced from the peak reached in 2004, they are not yet completely phased out (see figure 2A.1). Despite a favorable electricity generation mix (hydro versus thermal), public transfers to the energy sector still absorbed 0.4 percent of GDP in 2008.

Ghana is currently facing a severe power crisis that could have significant macroeconomic repercussions. The prevailing tariff structure does not enable the sector to be financially sustainable, and there are legitimate concerns about raising tariffs (World Bank 2011). Although electricity tariffs were substantially increased to eliminate subsidy burdens (see figure 2A.6) and quarterly power tariff reviews were introduced, energy pricing remains an area of risk (IMF 2011).

As a result of fuel reforms, some petroleum products such as gasoline and diesel are still taxed quite heavily. Combined tax elements for gasoline make up about 35 percent of the final pump price. Oil marketing companies are allowed to set their pump prices up to a certain level, but there is a ceiling on these prices established by the NPA. A small portion of the tax is also used to subsidize kerosene. In June 2008, the Ghanaian parliament enacted two laws to address consumer grievances about rapid fuel price increases. The higher fuel prices were also having indirect effects on other products such as food and transportation through their input-output links. One of the laws (the Debt Recovery Levy) included a reduction in some fuel tax elements on select petroleum products such as gasoline, diesel, and kerosene, while the other addressed import duties on food items. That law costs the government a 6 percent shortfall in full tax revenue. The combined effects of the two laws were mixed because the price reduction was hardly reflected in transportation and food items.

Reform Efforts

Fossil Fuels

- Petroleum product prices were increased in 2003 by around 90 percent (see figure 2A.3). The automatic adjustment formula was effectively abandoned in early 2003 when continued increases in world prices were not passed on to consumers (Coady et al. 2006).

- In 2004, the government launched a poverty and social impact assessment (PSIA) for fuel. Guided by a steering committee of stakeholders from ministries, academia, and the national oil company, the PSIA was completed in less than a year. By the time the government announced the 50 percent price increases in February 2005, it could use the PSIA findings to make its case to the public for liberalizing fuel prices (Coady et al. 2006).

• In mid-February 2005, when the Ghanaian government increased petroleum prices by 50 percent, it also announced its intention to introduce again an automatic adjustment formula. In addition, it emphasized its commitment to continue sectoral reforms that would further increase private sector participation in the import and distribution of petroleum products (Coady et al. 2006). The government of Ghana used budgetary savings to expand the existing rural electrification scheme. This was a prominent component of the expenditure package introduced simultaneously with the fuel price increases. The incidence of the benefits from these expenditures was found to be strongly progressive (Coady et al. 2006). The mitigation measures— transparent and easily monitored by society—included an immediate elimination of fees at government-run primary and junior secondary schools and a program to improve public transport. Extra funds were made available to an existing program, the Community Health Compound Scheme, to enhance primary health care in the poorest areas (Bacon and Kojima 2006).

• In June 2005, as previously mentioned, the government established the NPA to monitor the implementation of the pricing mechanism and facilitate the withdrawal of government from the politically sensitive issue of petroleum pricing. Prices increased again in June, August, and October of 2005 as a result of climbing world oil prices. In 2006, to keep up with frequent hikes in international prices of oil and to continue to control the short-term subsidies bill, quarterly price adjustments were replaced by monthly price adjustments (Coady et al. 2006). The minister of finance launched a public relations campaign via broadcast explaining the need for the price increases and announcing measures to mitigate their impact. In November 2007, prices were increased again by 35 percent (see figure 2A.3).

• In November 2007, to mitigate the effects on consumers from the continued price increases, 6 million energy-saving compact fluorescent lamps were distributed free of charge with the aim of reducing the electricity bills by up to 50 percent. A poverty alleviation program called Livelihood Empowerment against Poverty (LEAP) was also introduced to provide direct cash transfers to the poor and assist additional social services in supporting the poor against price increases. LEAP was extended from 35,000 to 55,000 people in 2012.

- Since the National Democratic Congress took office in January 2009, prices were increased again several times, with the latest being a 5 percent increase that took effect in October 2009.

- Petroleum pump prices were raised by 30 percent in January 2011 to cost-recovery levels after a rise in global prices in late 2010. Hedging operations conducted since October 2010 have provided protection from global price increases. With the benefits of hedging, pump prices were kept at current levels through mid-2011 (IMF 2011). By 2011, the bulk of TOR's bank liabilities were cleared through public debt issues. As a result, it has regained operational independence. Plans to modernize the refinery and strengthen its longer-term commercial viability have been developed and were to be shared with the World Bank for its assessment (IMF 2011).

Electricity
- The Ministry of Energy is responsible for energy policy formulation and implementation, while the Energy Commission, set up under Act 541 in 1997, is responsible for energy policy and strategy advice, national energy planning, licensing, and technical regulations. The Public Utilities Regulatory Commission (PURC), set up under Act 538 in 1997, regulates electricity tariffs and customer services. The electricity generation and transmission functions lie with the Volta River Authority, while electricity distribution in the southern part of the country is the responsibility of the Electricity Company of Ghana, and the Northern Electricity Department is responsible for distribution in northern Ghana (Estache and Vagliasindi 2007).

- Ghana has a long history of attempts to reconcile cost-reflective tariffs and affordability. Tariff adjustments to reflect changes in cost (such as exchange rate or inflation) were implemented back in 1994 and 1997. The first intense opposition to tariff increases by several industrial associations—including the Civil Servants Association, the Association of Ghana Industries, and the Trades Union Congress—took place in May 1997 subsequent to an increase in electricity tariffs of over 300 percent. Following this episode, the president suspended the increase, and draft legislation was enacted to establish PURC as an independent regulatory agency (Estache and Vagliasindi 2007).

- Since its establishment in October 1997, PURC has adjusted electricity tariffs six times. The adjustments were made in 1998 (February and September), 1999, 2001 (May), 2002 (August), and 2003 (March). The first major electricity tariff increase was over 400 percent for all categories of consumers, and the second increase was 103 percent in 2001. The combined increase in 2002 and 2003 was 72 percent. Two further adjustments in January and April 2004 have since been implemented.

- Among the more recent additional tariff increases, one implemented in June 2010 implied increases of up to 130 percent for nonresidential users as well as an increase for the average tariff.

Poverty Alleviation Measures

Social Safety Nets

The best-targeted program appears to be LEAP, a program designed to provide cash transfers to households in extreme poverty. LEAP aims to reach the poorest of the poor, defined by the program as the bottom 20 percent of the poor. The data suggest that three-fourths of the transfers provided by LEAP reach the bottom two income quintiles of the population, and the share reaching the poor is estimated at 57.5 percent (see figure 2A.13). An expansion of the program would thus generate substantial benefits for the poor and would also help in reducing the share of program costs currently devoted to administration and delivery (World Bank 2011).

Another program that appears to be well targeted is the indigent exemption for the registration and coverage of very poor households under the National Health Insurance Scheme (NHIS). The share of NHIS outlays benefiting the poor is estimated above 50 percent. Given low levels of enrollment under this exemption today compared with the share of the population in extreme poverty, districts should be encouraged to make more extensive use of the indigent exemption. Other programs include general funding for primary education (whose share of outlays benefiting the poor is estimated at 32.2 percent); general funding for health services (30.8 percent); and free maternal (ante- and postnatal) and child care (29.1 percent) (World Bank 2011).

Fossil Fuels

Unsurprisingly, the targeting of fuel subsidies is poor. Almost any universal consumption subsidy will disproportionately benefit the rich because they, by definition, account for a relatively high proportion of total income and consumption. Even an equal uniform transfer to all households would be better targeted than existing subsidies because 40 percent of benefits would accrue to the poorest 40 percent of households (Coady et al. 2006).

It is estimated that only 2.9 percent of the volume of subsidies for diesel and gasoline reach the poor. Given that oil products are used as intermediary inputs for a wide range of activities (transportation, for example), the share of the subsidies that indirectly reach the poor is likely to be slightly higher (World Bank 2011). Gasoline is consumed primarily by higher-income households, whereas kerosene is relatively more important in the budgets of lower-income households (see figure 2A.11).

Kerosene subsidies are more progressive because kerosene is the dominant component of the energy budget for lower-income households, and it accounts for over 67 percent of all household energy expenditures; 20.7 percent of kerosene subsidies reach the poor and protect them against the fluctuation of oil prices (see figure 2A.13). The benefit is also progressive because kerosene is the only oil-related product that is consumed in a substantial way by the poor (World Bank 2011).

Del Granado et al. (2010) consider the direct and indirect impacts of a US$0.25 per liter increase in fuel prices in the case of Ghana. The *direct* impact of phasing out subsidies—considering only the impact on the consumption of fuels for cooking, heating, lighting, and private transport—is a loss of 5.6 percent of real income, mostly due to the reduction in kerosene consumption. The *indirect* impact—through higher prices for other goods and services—is twice as high as the direct impact, accounting for almost 12 percent of real income (see figure 2A.12).

Electricity

Ghana, as many other developing countries, opted for lifeline tariffs as the means to reduce the cost of energy supplies. Lifeline tariffs usually come in two- or three-block versions. Two-block lifeline tariffs have a lower tariff for energy consumed up to a certain limit, usually set quite low at a level of minimal, or lifeline, energy consumption. Three-block lifeline tariffs introduce a third, higher tariff for energy consumed over a certain limit to discourage very high levels of use (which may be a sign of inefficiency) and to encourage fuel switching. The latter is particularly important in cases where electricity is used for heating and cheaper, more efficient alternatives, such as gas, may be available. Ghana's residential tariff structure was originally based on five blocks, defined according to the level of consumption.

The five blocks were subsequently reduced to four and then to three in May 2001 and March 2003, respectively, during the tariff review process. The lowest block offers a flat rate to customers with consumption equal to or below 50 kilowatt-hours (kWh) per month. The lifeline tariff has been a part of the tariff system for more than a decade and was originally created to minimize the cost of billing small accounts. It became partially subsidized by the government only since August 2002, with the government paying a subsidy of 5,000 cedis to the electricity supplier per lifeline customer (the lifeline being equal to 9,000 cedis). When the automatic adjustment formula was set to start in 2002, the government increased the subsidy to 6,080 cedis to protect this block from the tariff increase. However, the adjustment with the formula did not take place until October 2003 (World Bank 2005).

After the November 2007 tariff increase, the government of Ghana was worried that certain consumers would be unable to afford electricity at this new rate. So the lifeline consumption threshold was increased to include customers in the 51–150 kWh consumption block. The current block tariff has been in force since June 2010 (see figure 2A.8). Because the lifeline threshold drops to 50 kWh in the new tariff regulation, consumers in the 51–150 kWh block faced a steep 79 percent price increase (World Bank 2010).

An early assessment of the targeting effectiveness of the lifeline tariff for the fiscal year 2005/06 showed that only 8 percent of the subsidies for those who consumed lower amounts of electricity reached the poor. Changes in tariff structure since 2005/06 may have increased the share of benefits accruing to the poor, but targeting performance is likely to remain poor because many residential electricity customers who benefit from the lifeline are nonpoor, given the electricity access divide between the lowest and top deciles of the population (see figure 2A.9).

One of the shortcomings of the lifeline subsidy mechanism is that the benefits do not cover those residents who live in compound houses. Those individual families would be better off with meters because compound houses certainly use more than 50 kWh per month, but since they pay the bill collectively, it ends up being too high to be eligible for the subsidy (ESMAP 2005). Additional evidence from household survey data from 2003 shows that those Ghanaians who fall under the poverty line do not tend to have access to electricity in rural areas, where the majority of the poor are concentrated. Only 7 percent of rural poor people use electricity for lighting, while 93 percent use kerosene. Roughly 54 percent of the urban poor use electricity for lighting (see figures 2A.10 and 2A.11) (ESMAP 2005).

A more recent study (World Bank 2010) shows that there is no change in the poverty levels of users in the 10th and 20th income percentiles, or the extremely poor. Conversely, under conservative simulations, consumers whose income is just slightly above the poverty line—those at the lower-middle income level—might be adversely affected. These calculations do not factor in income growth or substitution of other energy sources and, therefore, present a worst-case scenario. In this case, poverty levels for those in the 30th and 40th income percentiles increase by 0.7 and 2.7 percentage points, respectively, and the overall poverty rate increases by only 0.3 percentage points. These income deciles are the most affected by the new tariffs because the majority of the electricity users in these income brackets are in the 51–150 kWh consumption bracket.

Household survey data suggest that providing connection instead of consumption subsidies could substantially improve targeting, but providing such connection subsidies supposes also that cost recovery is adequate in order not to increase sector deficits further.

Grid extension to rural areas of the north also receives government subsidies in the form of inputs as part of the Program on Rural Electrification, as does the program on kerosene distribution for rural areas (Kankam and Boon 2009). Policy makers are hoping for increased penetration of renewable energy in the rural energy mix, in which independent power producers would play a key role and either supply more to the national grid for those who have grid access or provide off-grid electrification in remote areas and in those areas where the share of new investment for electricity is declining. The Global Partnership on Output-Based Aid (GPOBA) is currently implementing a project on installing solar photovoltaic (PV) systems in precisely those areas where the poverty rates are highest, and to which the power grid is not anticipated to reach for at least another 10 years. The project aims to install 15,000 solar lanterns and solar home systems to benefit 90,000 people. This is three times the number of PV systems currently installed in Ghana. GPOBA is contributing 50 percent of the total costs of supply, installation, maintenance, and battery replacement while consumers pay the other 50 percent—10 percent as down payment and 90 percent as loans (Mumssen, Johannes, and Kumar 2010).

Key Lessons Learned

Ghana has made significant progress in reducing fossil fuel subsidies. However, observers are worried that with Ghana's planned increase in oil production and its potential for high oil exports, the government will not be able to resist the pressure to reinstate subsidies and will backtrack on the reforms. As we know from other cases, a number of political factors such as elections (which in Ghana were scheduled for 2012) and an improved fiscal situation may affect decisions on subsidies.

Ghana offers many valuable lessons in subsidy reform. The subsidy reform and the palliative measures were preceded by widespread media and information campaigns about the goals of the reform as well as the planned benefits designed to lessen the impact on those who would be most affected. The automatic adjustment formula and the establishment of the NPA for the government to remove itself from pricing decisions are moves that for many signify more credibility and less political interference.

Annex 2.1 Ghana Case Study Figures

INCOME LEVEL: Low income
REGION: Sub-Saharan Africa
ENERGY NET IMPORTER/EXPORTER: Net importer
SUBSIDIES: Kerosene, electricity
PHASING OUT SUBSIDIES: Successful

Fiscal Burden of Energy Subsidy in Ghana

Figure 2A.1   Explicit Budgetary Energy Subsidies in Ghana, 2003–10

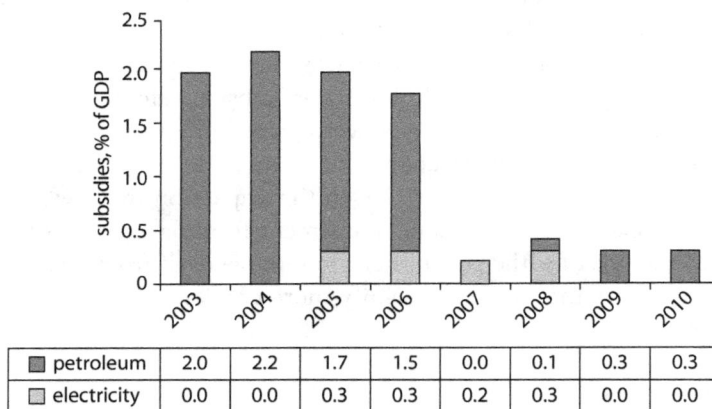

	2003	2004	2005	2006	2007	2008	2009	2010
petroleum	2.0	2.2	1.7	1.5	0.0	0.1	0.3	0.3
electricity	0.0	0.0	0.3	0.3	0.2	0.3	0.0	0.0

Sources: IMF staff reports, various years.
Note: Budgetary electricity subsidies equal zero after 2008.

Figure 2A.2   Implicit Subsidies of the Power Sector in Ghana, 2004–09

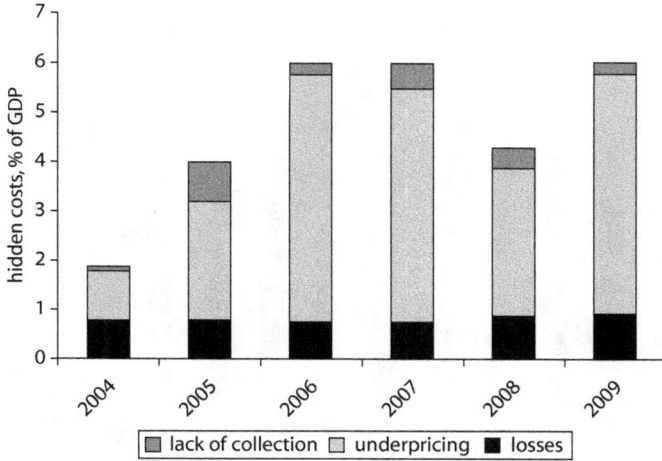

Source: AICD n.d.

Note: Implicit subsidies (or hidden costs) are defined as the difference between actual receipts and the revenue that the energy company (for example, a utility involved in the distribution of electric and natural gas) would receive were it to be in operation with cost-recovery tariffs based on efficient operation with normal losses and with full bill collection.

Fuel Prices and Road Sector Consumption in Ghana

Figure 2A.3   Domestic Retail Fuel Prices in Ghana, 2002–10

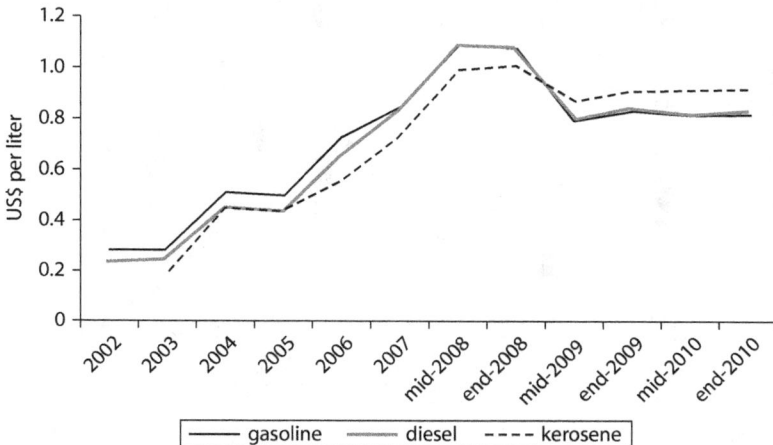

Sources: Elaboration of data from GIZ n.d.; IMF 2010; and additional data from individual country information.

Figure 2A.4   Road Sector Diesel Consumption in Ghana, 1998–2008

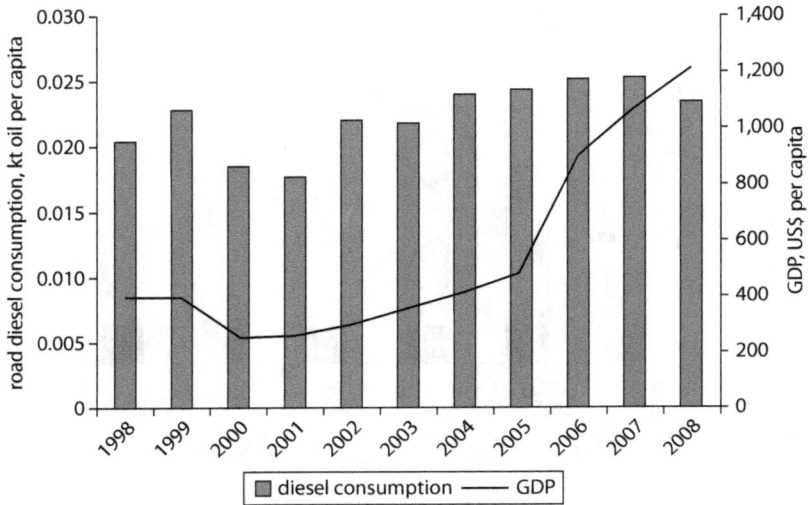

Source: World Bank, World Development Indicators.

Figure 2A.5   Road Sector Gasoline Consumption in Ghana, 1998–2008

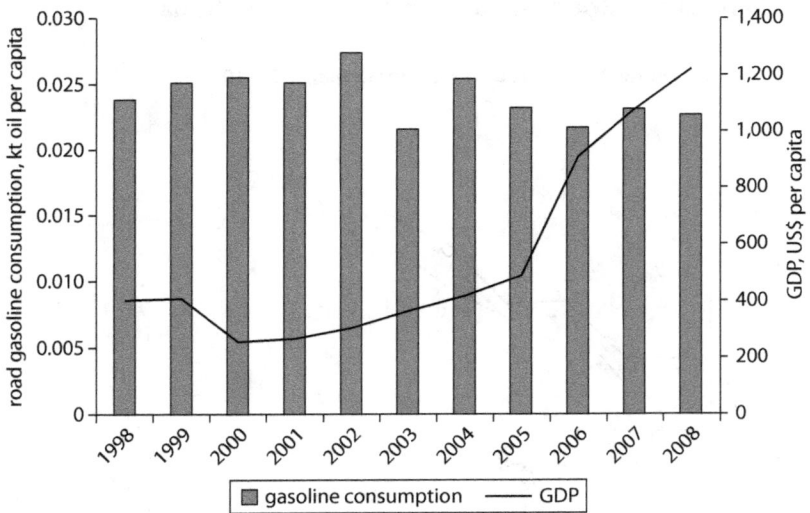

Source: World Bank, World Development Indicators.

Electricity Price and Power Consumption in Ghana

Figure 2A.6   Electricity Price in Ghana, 2003–09

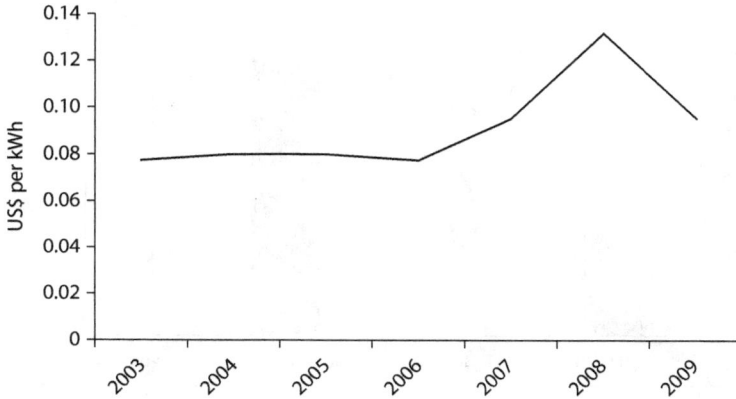

Source: PURC.

Figure 2A.7   Power Consumption Per Capita in Ghana, 1998–2008

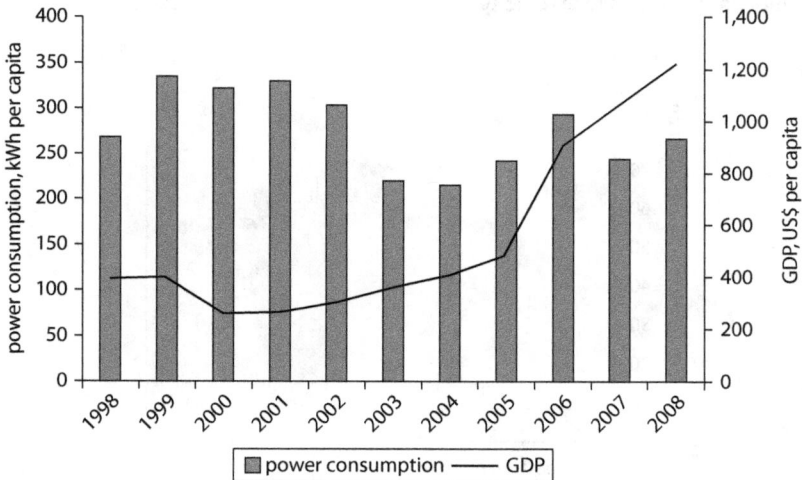

Source: World Bank, World Development Indicators.

Poverty Impact Evidence from Household Surveys in Ghana

Figure 2A.8   Electricity Block Tariffs in Ghana, 2010

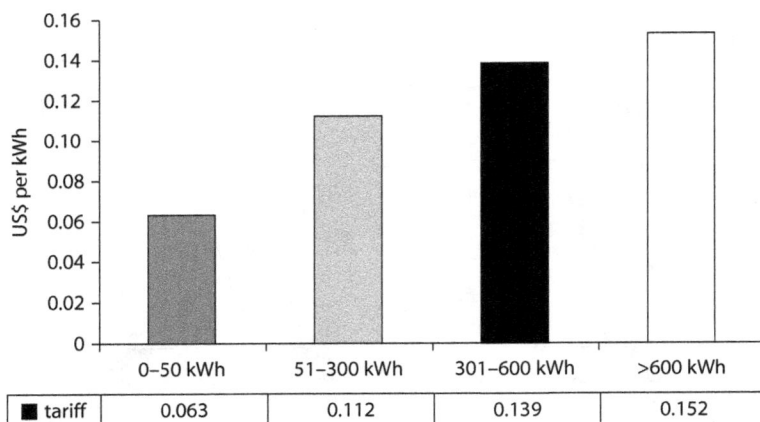

	0–50 kWh	51–300 kWh	301–600 kWh	>600 kWh
■ tariff	0.063	0.112	0.139	0.152

Source: PURC.
Note: Effective June, 1, 2010.

Figure 2A.9   Access to Electricity in Ghana, by Income Quintile, 2011

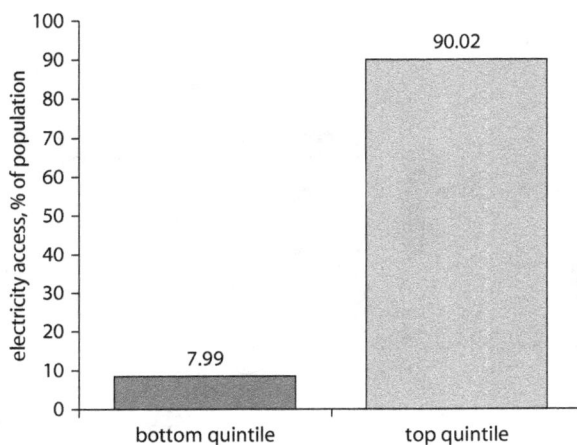

Source: AICD n.d.

Figure 2A.10   Use of Energy Sources in Ghana, by Rural and Urban Population, 2011

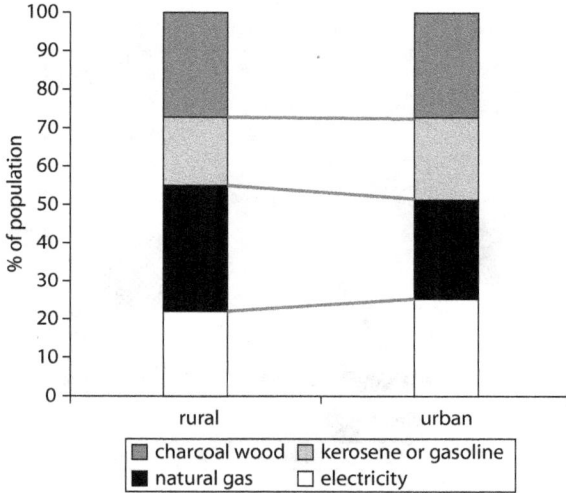

Source: AICD n.d.

Figure 2A.11   Energy Expenditure in Ghana, by Income Quintile, 1999

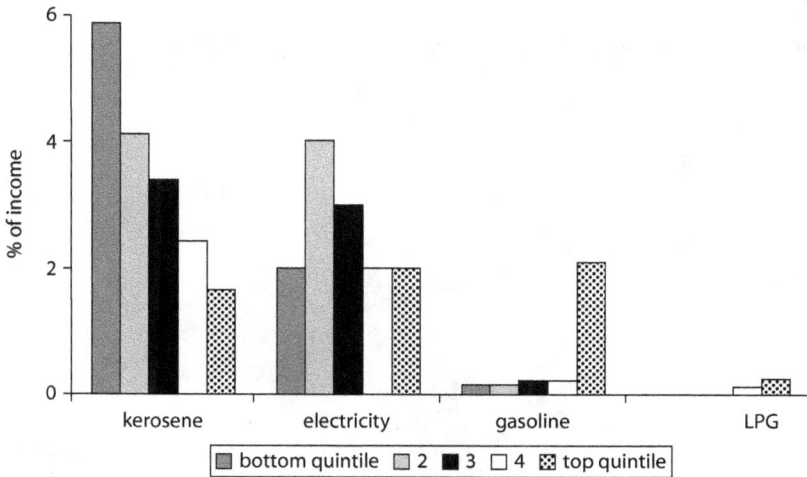

Source: Del Granado, Coady, and Gillingham 2010, based on 1999 household expenditure survey.
Note: LPG = liquefied petroleum gas.

Figure 2A.12   Welfare Impact of Removing Fossil Fuel Subsidies in Ghana, 1999

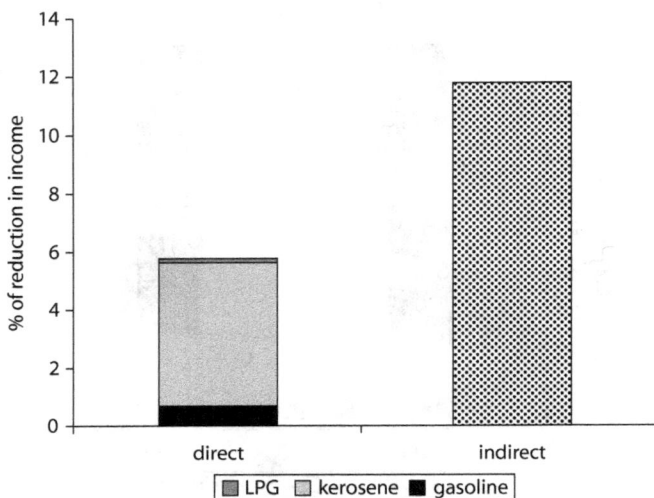

Source: Del Granado, Coady, and Gillingham 2010, based on 1999 household expenditure survey.
Note: LPG = liquefied petroleum gas. Indirect impacts include higher prices for non-energy goods and services consumed by households to the extent that increased production costs and consumer prices reflect higher fuel costs.

Figure 2A.13   Targeting of Social Programs and Energy Subsidies in Ghana, 2011

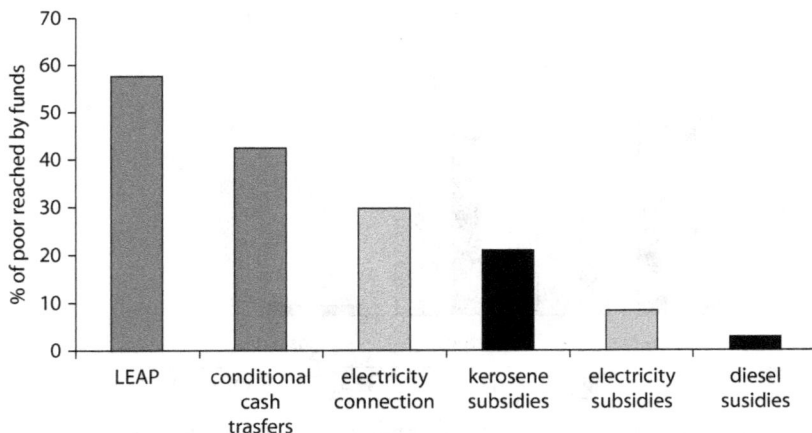

Source: World Bank 2011.
Note: LEAP = Livelihood Empowerment against Poverty.

References

AICD (Africa Infrastructure Country Diagnostic) database. n.d. World Bank, Washington, DC. http://www.infrastructureafrica.org/tools/data.

Bacon, Robert, and Masami Kojima. 2006. "Coping with Higher Oil Prices." Report 323/06, Energy Sector Management Assistance Program and World Bank, Washington, DC.

Coady, David, Moataz El-Said, Robert Gillingham, Kangni Kpodar, Paolo Medas, and David Newhouse. 2006. "The Magnitude and Distribution of Fuel Subsidies: Evidence from Bolivia, Ghana, Jordan, Mali, and Sri Lanka." Working Paper 06/247, International Monetary Fund, Washington, DC.

Del Granado, J., D. Coady, and R. Gillingham. 2010. "The Unequal Benefits of Fuel Subsidies: A Review of Evidence for Developing Countries." Working Paper 10/202, International Monetary Fund, Washington, DC.

ESMAP (Energy Sector Management Assistance Program). 2005. "Ghana: Poverty and Social Impact Analysis of Electricity Tariffs." Technical Paper 88, ESMAP, World Bank, Washington, DC.

Estache, Antonio, and Maria Vagliasindi. 2007. "Infrastructure for Accelerated Growth for Ghana: Investment, Policies and Institutions." In *Ghana: Meeting the Challenge of Accelerated and Shared Growth.* Country Economic Memorandum, Report 40934-GH, Vol. 3, World Bank, Washington, DC.

GIZ (German Agency for International Cooperation). n.d. International Fuel Prices database. GIZ (formerly GTZ), Bonn. http://www.gtz.de/en/themen/29957.htm.

IMF (International Monetary Fund). 2010. Retail domestic fuel prices data sheet, IMF, Washington, DC. http://www.imf.org/external/pubs/ft/spn/2010/data/spn1005.csv.

———. 2011. "Ghana: 2011 Article IV Consultation and Third and Fourth Reviews Under the Arrangement Under the Extended Credit Facility." Country Report 11/128, IMF, Washington, DC.

Kankam, Stephen, and Emmanuel K. Boon. 2009. "Energy Delivery and Utilization for Rural Development: Lessons from Northern Ghana." *Energy for Sustainable Development* 13 (3): 212–18.

Mumssen, Yogita, Lars Johannes, and Geeta Kumar. 2010. *Output-Based Aid: Lessons Learned and Best Practices.* Directions in Development Series. Washington, DC: World Bank.

World Bank. 2005. "Ghana Energy Policy Economic and Sector Work Papers: The Electricity Sector." Internal reports, World Bank, Washington, DC.

———. 2009. "Economy-Wide Impact of Oil Discovery in Ghana." Report 47321-GH, World Bank, Washington, DC.

————. 2010. "Ghana: Poverty and Social Impact Analysis (PSIA): Electricity Tariffs." Government of Ghana and World Bank, Washington, DC.

————. 2011. "Republic of Ghana: Improving the Targeting of Social Programs." Report 55578-GH, World Bank, Washington, DC.

CHAPTER 3

India

Incentives to Energy Subsidy Reforms

India stands fifth in the world in terms of electricity generation capacity. However, it still faces substantial energy and peak deficits. The country's power sector is characterized by a wide demand-supply gap, dominant presence of state-owned utilities, unelectrified rural areas, a need for tariff rationalization to address cross-subsidies, and weak financial health of state-level utilities (World Bank 2010). As of 2010, in India, close to 64 percent of total installed capacity was from thermal sources (predominantly coal); hydropower contributed 23 percent to the total installed capacity; and the balance (13 percent) came from nuclear and renewable energy sources.

When the central government budget deficit reached almost 7 percent of gross domestic product (GDP) in 2009, the government started considering a new fuel pricing policy for gasoline, diesel, liquefied petroleum gas (LPG), and kerosene. A number of reforms were implemented in 2010 with the full liberalization of gasoline prices and the increase in prices of all petroleum fuels.

To finance the underrecovery of oil marketing companies (OMGs), the government employs a number of different mechanisms. A small percentage is financed through on-budget subsidies, which have accounted for 0.1 percent of GDP (see figure 3A.1). The companies are obliged to sell at the prices that are determined by the central government, with the exception of gasoline, which was liberalized in 2010. These prices are below the cost of buying and distributing products. The difference between the OMGs' costs and their revenues (underrecovery) represents a subsidy to consumers.

However, the majority of financing takes place off-budget. The important source of financing was, until 2008, the issuance of oil bonds to the OMGs. Another portion of underrecoveries is financed by oil and gas producers, which are required by the government to transfer shares of their profits to the OMGs. The implicit subsidies through government off-budget mechanisms reached a peak of about 1.3 percent of GDP in 2008 (see figure 3A.2). Although in 2009 and 2010 the amount of subsidies decreased substantially—to an insignificant amount because of the decrease in the world prices of oil—in 2011 the amount of subsidies went up again (OECD 2011). In the 2011 budget, the government announced plans to stop issuing oil bonds and switch to directly subsidizing the OMGs instead.

Reform Efforts

Fossil Fuels
Under the Administrative Pricing Mechanism (APM) between 1976 and 2002, petroleum product prices were fixed by the government. Kerosene and LPG were cross-subsidized by higher-priced gasoline, diesel, and other products. In 2002, the APM was dismantled by the government to give oil companies some freedom to sell products at market prices while also announcing that the subsidies for residential kerosene and LPG would be phased out in the subsequent three years.

Several attempts to address kerosene subsidies were made in the past, none of them fully successful. The government tried using blue coloration for subsidized kerosene twice—in the 1980s and again in 2006—to distinguish it from nonsubsidized kerosene and reduce its profitability in the black market, but pump owners and other rent seekers in the black market found ways to neutralize the blue dye using clay and other coloring. The program was suspended in 2008. A coupon system for supplying subsidized kerosene was also attempted to ration the quantity of kerosene distributed, but the scheme never took off because of political resistance by kerosene dealers and lobbyists. Smart cards were another rationing scheme proposed in 2005 and 2007; however, several Indian states rejected the proposal (Shenoy 2010).

Three OMGs dominate India's retail market for petroleum products because privately owned companies are not allowed to receive subsidies from the government. In June 2010, gasoline prices were fully liberalized, with immediate effect. OMGs can now set their own retail prices of gasoline—after the government's approval. The government also announced its intention to follow up by liberalizing diesel prices. At the same time, wholesale prices for kerosene and LNG were raised by 33 percent and 11 percent, respectively, while the price of diesel was raised by 5 percent, though the government reserved the right to reverse these measures if international oil prices climbed to excessive levels (see figure 3A.3).

Electricity

In June 2003, the Electricity Act was introduced. The enactment of the Act paved the way to undertake comprehensive market reforms, including mandatory unbundling, the creation of independent regulatory commissions, multiyear tariff approvals, extension of the Availability-Based Tariff within the state, compulsory metering, and the declaration of electricity theft as a criminal offense.

The Tariff Policy (introduced in 2006) provides guidelines to regulators for fixing tariffs for generation, transmission, and distribution. Among other things, it has made it mandatory for distribution licensees to procure power through competitive bidding (except in cases where a state-owned company has been identified as the developer). To the end of improving the competitiveness of industrial and commercial tariffs, the Tariff Policy suggests bringing down the cross-subsidy progressively at a linear rate, to a maximum of 20 percent of its opening level (World Bank 2010).

The retail consumer tariffs are not reflective of the cost of service. The industrial tariff is still high, and the agricultural tariff (accounting for about 25 percent of consumption) is well below the cost of service, as are the residential or domestic tariffs (see figure 3A.6). The number of hours and the quality of supply are low, particularly in rural areas, and there is planned and unplanned load shedding in various consuming areas. Even states that are at a more advanced stage of electricity reform continue to report significant problems in unmetered supply (World Bank 2010).

Poverty Alleviation Measures

Evidence from Household Surveys

The cross-subsidies embedded in the tariff structure that benefit households consuming low levels of electricity are highly regressive. (Figure 3A.8 displays the inverted block tariff for the New Delhi Municipal Council.) Analysis of the benefit incidence of electricity subsidies, based on the 2002 India National Sample Survey, shows that the top income quintile captures 30–45 percent of the benefits (see figure 3A.9) because the bottom quintiles are largely not connected to the grid.

Major changes are evident in the distribution of the population using different energy types across rural and urban households from 1983 to 2005 (see figure 3A.10). The percentage of the population using LPG increased from 9 percent to 61 percent in urban areas. At the same time, in rural households, the uptake of LPG was much slower, and even in 2004–05 only 12 percent of the rural population used this fuel. Electricity access also changed dramatically over this period. Whereas 15 percent of the rural population and 58 percent of the urban population were using electricity in 1983, by 2005, 54 percent of the rural and 91 percent of the urban population were doing so. Kerosene is used widely by all households. Although the percentage of rural population using this fuel has not changed much over this period, the percentage of the urban population using kerosene declined from 92 percent to 55 percent. Although the share of traditional biomass energy users (both firewood and dung) in rural areas remained relatively constant during this period, the percentage of the population using traditional biomass energy in urban areas was halved (Pachauri and Leiwen 2008).

A comparison of energy choices in households across rural and urban deciles provides further evidence that in 2005, rural households still lacked access to electricity and relied on biomass, whereas among urban households, only 20 percent of the households belonging to the top quintile relied on biomass and 30 percent on kerosene (see figure 3A.11). The finding suggests that subsidies for modern fossil fuels are biased toward the urban sector and that rural households continue to use biomass as a primary fuel source despite subsidies in place to provide incentives to switch out of biomass. Biomass is still making up the lion's share of rural fuel consumption for cooking, LPG being the dominant fuel for this purpose in urban areas, as figure 3A.11 shows.

The share of expenditure in biomass and kerosene by higher quintiles tends to fall significantly, while the use of these fuels is significantly higher for the poorest households. On the contrary, the share of expenditure in electricity—but particularly for LPG, gasoline, and diesel—tends to increase significantly by higher quintiles, while the use is significantly lower for the poorest households (see figure 3A.12).

The use of kerosene subsidies represented a very expensive way to protect poor households, as about half of the subsidized kerosene supplies are diverted illegally to arbitrage the price difference between the subsidized rate and the market rate and is thus not always available to the poor. A similar black market exists for LPG, where the fuel is sold at market rates for industrial and commercial purposes. Half of government supplies never reached households, and this leakage cost the government close to US$1 billion in fiscal year 2000 (Bacon and Kojima 2006). Using a 2009 tax and price regime with crude oil prices of US$70 per barrel, Shenoy (2010) estimates that as much as US$1.6 billion in black market money was collected as a result of diverting subsidized kerosene, while diversion of kerosene to gasoline in the form of fuel adulteration brought profits of about US$2.7 billion. Similarly, diversion of LPG to nonresidential sectors amounted to about US$1.2 billion (Shenoy 2010).

Because LPG is mostly used by higher-income groups and those who switch to LPG are those with growing household incomes, the subsidy is regressive and unlikely to have any effect in displacing biomass. Another problem with the use of subsidies is that households were not always entitled to subsidized kerosene because many did not possess permanent addresses. This can be the case for migrant households that undertake seasonal migration (Gangopadhyay, Ramaswami, and Wadhwa 2005).

Social Safety Nets
The main program of social assistance to the poor in India is the Public Distribution System. The system uses a below-poverty line (BPL) as a household targeting mechanism, but it is regarded by experts as suboptimal and with a poorly designed proxy means testing. A World Bank study on social protection in India acknowledges the progressivity of BPL cards but identifies substantial inclusion errors in BPL card holding, with around 28 percent of all BPL cards nationally held by households in the top 40 percent of the distribution (World Bank 2007).

In 2011, the government announced the initiative to replace the existing mechanism of subsidies on kerosene and LPG with direct cash subsidies to people whose incomes are below the poverty line. The switch to the direct subsidies should help benefits go directly to the targeted groups and solve problems of fuel adulteration and speculation of subsidized fuels on the black market. The most obvious problem is identification of people below the poverty line. The introduction of the Unique Identification Authority of India to the program will be of crucial importance. The authority is responsible for implementing a project that will provide all households with a unique identification number. The program started with pilot projects in a number of states before the program was scheduled to be implemented in 2012.

Key Lessons Learned

The Indian government is well aware of the high costs imposed by energy subsidies and has been actively reviewing its energy pricing policies. Simulations in the Ministry of Petroleum and Natural Gas's Expert Group report on petroleum product pricing show that if the consumption of petroleum products grows at the same rate as in recent years, the current subsidy regime will result in underrecoveries in 2025 of US$88 billion for global crude prices at US$120 per barrel (Government of India 2010).

Many countries have attempted to shift from biomass use to promote cleaner fuels by using subsidies. Such shifts may be successful and may reduce negative externalities such as deforestation and air pollution, but if the subsidies are not transitory, the use of the promoted fuel at the subsidized price matures and becomes extremely difficult to change. In the case of India, kerosene and LPG subsidies have been unsuccessful in substituting for biomass and have thus failed to serve even the immediate purpose for which they were intended, indicating a more urgent need for reform.

Annex 3.1 India Case Study Figures

INCOME LEVEL: Lower-middle income
REGION: South Asia
ENERGY NET IMPORTER/EXPORTER: Net importer
SUBSIDIES: LPG, kerosene
PHASING OUT SUBSIDIES: Ongoing

Fiscal Burden of Energy Subsidy in India

Figure 3A.1   Explicit Budgetary Energy Subsidies in India, 2000–10

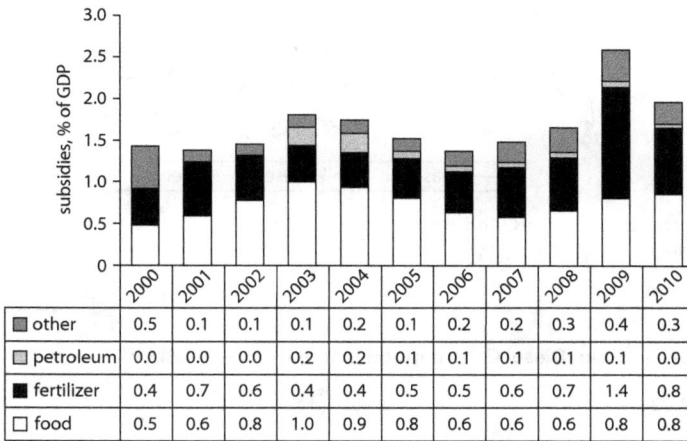

	2000	2001	2002	2003	2004	2005	2006	2007	2008	2009	2010
▨ other	0.5	0.1	0.1	0.1	0.2	0.1	0.2	0.2	0.3	0.4	0.3
☐ petroleum	0.0	0.0	0.0	0.2	0.2	0.1	0.1	0.1	0.1	0.1	0.0
■ fertilizer	0.4	0.7	0.6	0.4	0.4	0.5	0.5	0.6	0.7	1.4	0.8
☐ food	0.5	0.6	0.8	1.0	0.9	0.8	0.6	0.6	0.6	0.8	0.8

Source: Ministry of Finance, India Public Finance Statistics.

Figure 3A.2   Implicit Energy Subsidies in India, 2004–10

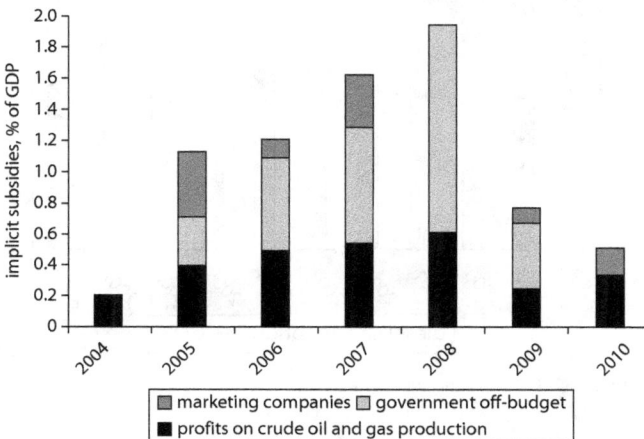

marketing companies ☐ government off-budget
■ profits on crude oil and gas production

Source: Ministry of Petroleum and Natural Gas, ICRA.

Fuel Prices and Road Sector Consumption in India

Figure 3A.3   Domestic Retail Fuel Prices in India, 2002–10

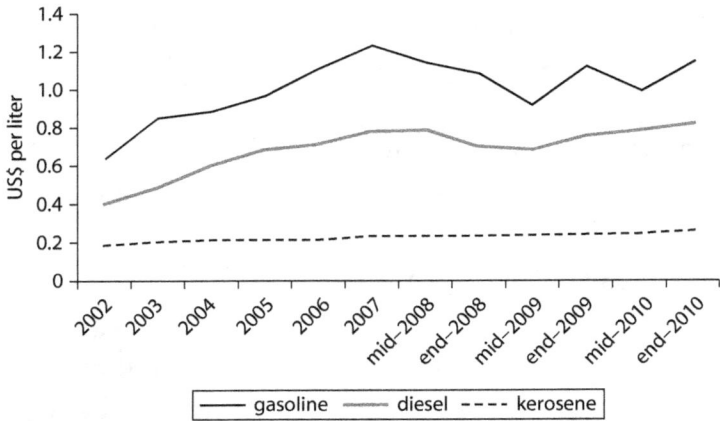

Source: Elaboration of data from GIZ n.d.; IMF 2010; and additional data from individual country information.

Figure 3A.4   Road Diesel Consumption in India, 1998–2008

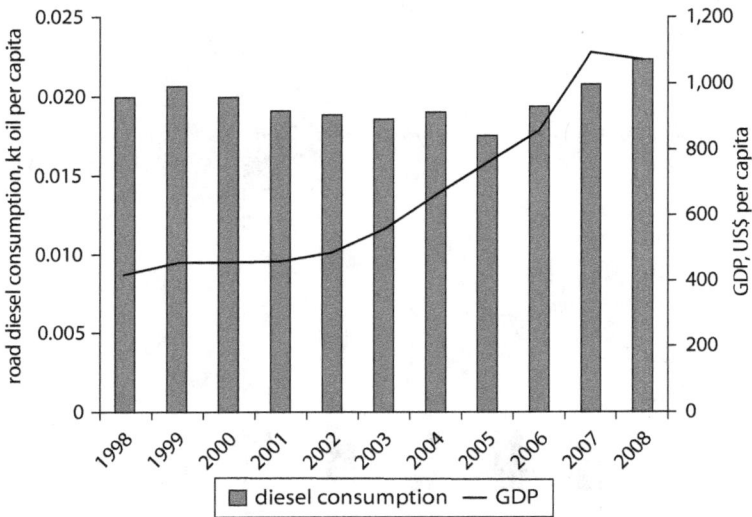

Source: World Bank, World Development Indicators.

Figure 3A.5   Road Gasoline Consumption in India, 1998–2008

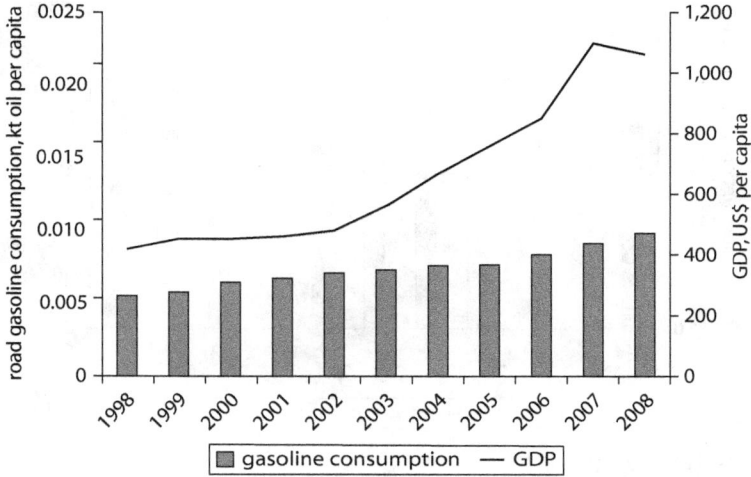

Source: World Bank, World Development Indicators.

Electricity Price and Power Consumption in India

Figure 3A.6   Electricity Price in India, 1998–2009

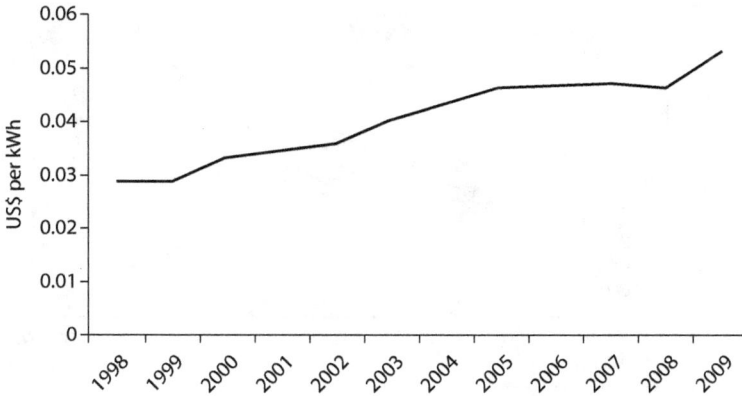

Source: IEA n.d.
Note: kWh = kilowatt-hour.

Figure 3A.7   Power Consumption Per Capita in India, 1998–2008

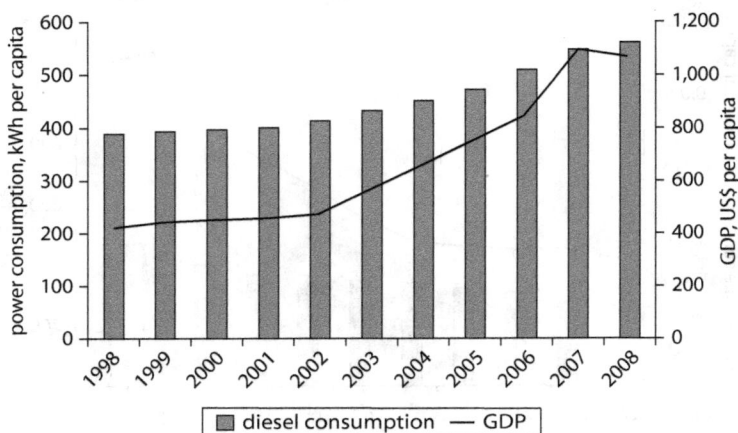

Source: World Bank, World Development Indicators.
Note: kWh = kilowatt-hours.

Poverty Impact Evidence from Household Surveys in India

Figure 3A.8   Electricity Block Tariffs in India, 2011

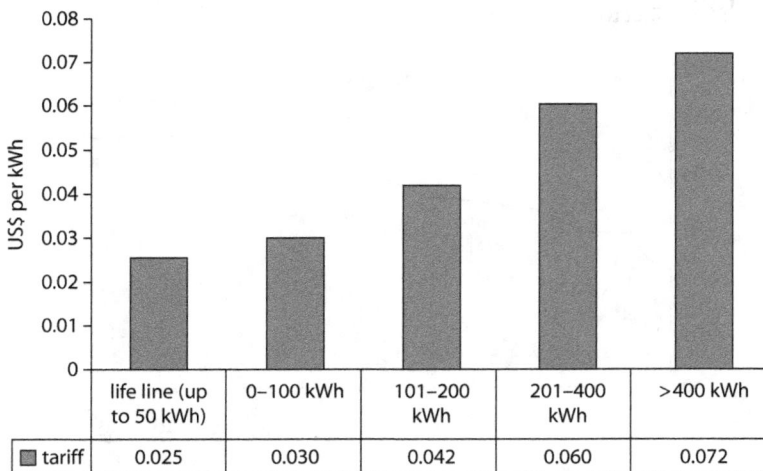

	life line (up to 50 kWh)	0–100 kWh	101–200 kWh	201–400 kWh	>400 kWh
tariff	0.025	0.030	0.042	0.060	0.072

Source: New Delhi Municipal Council.
Note: kWh = kilowatt-hour.

Figure 3A.9   Electricity Subsidy Benefit Incidence in India, by Income Quintile and State, 2002

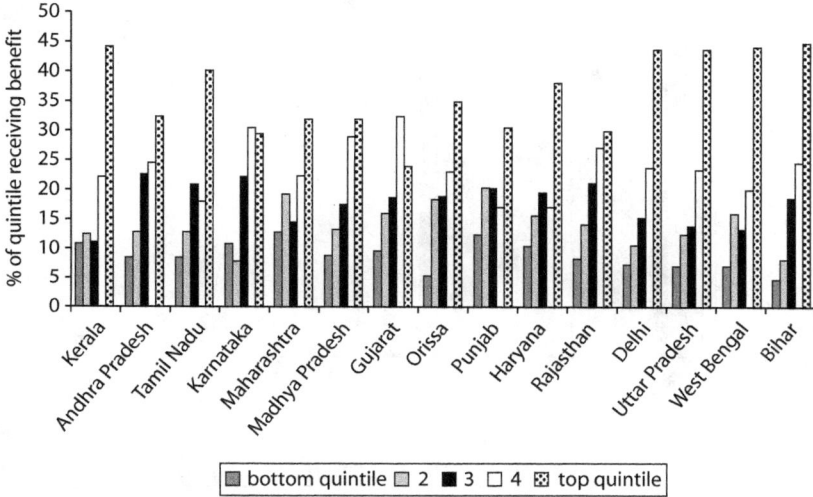

Figure 3A.10   Household Use of Energy Sources in India, 1983–2005

a. Rural

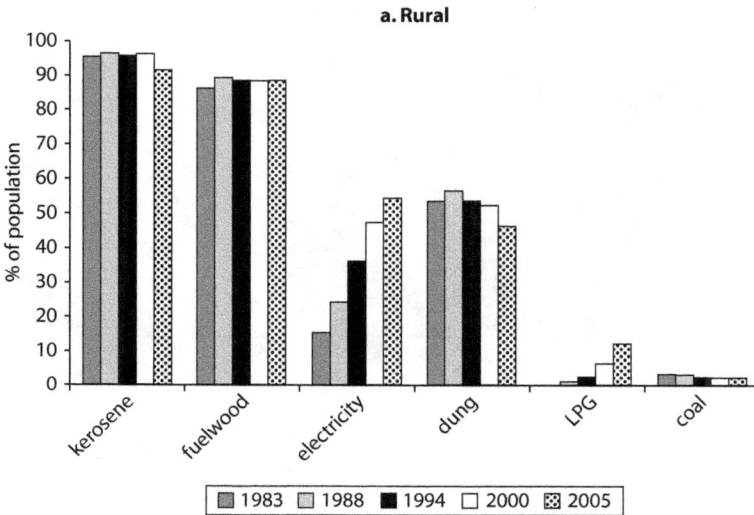

(continued next page)

Figure 3A.10   *(continued)*

b. Urban

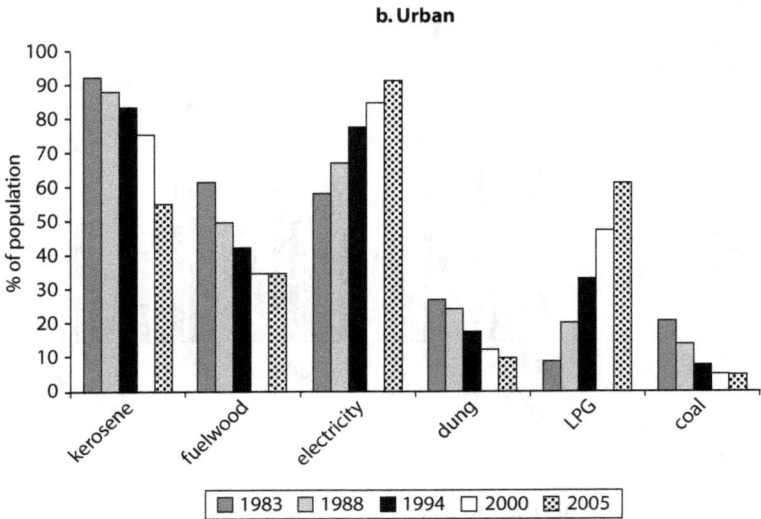

Source: NSSO 2007.
Note: LPG = liquefied petroleum gas.

Figure 3A.11   Household Use of Energy Sources in India, by Income Quintile, 2005

a. Rural

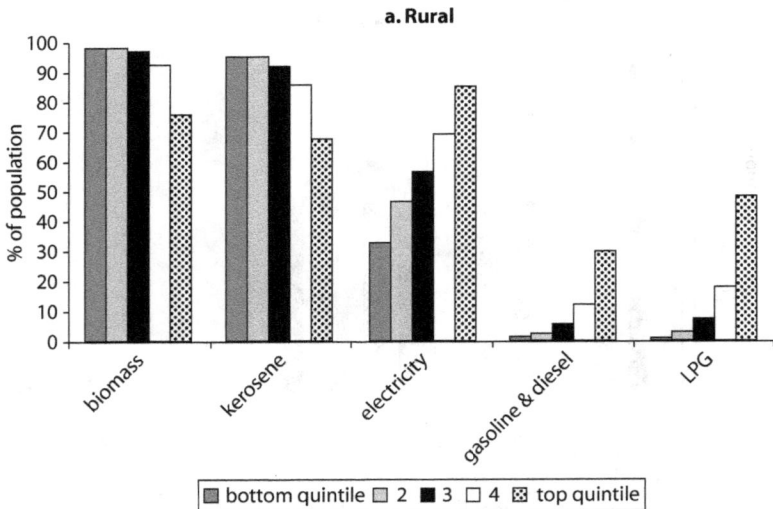

(continued next page)

Figure 3A.11   *(continued)*

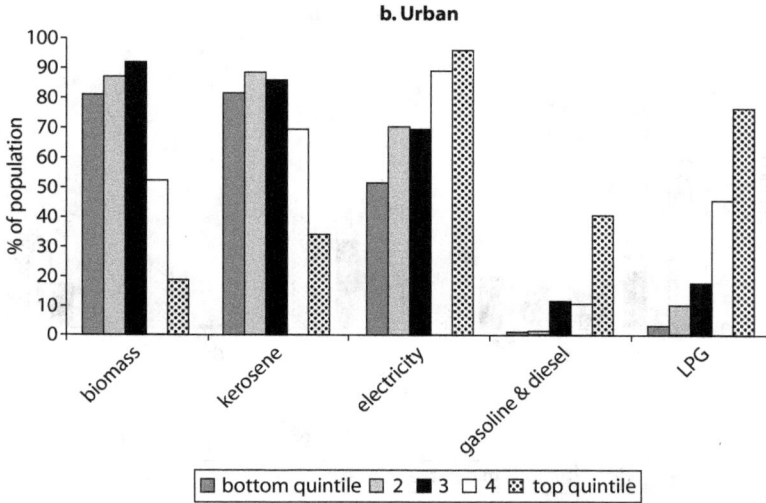

b. Urban

Source: NSSO 2007.
Note: LPG = liquefied petroleum gas.

Figure 3A.12   Household Energy Expenditure in India, by Income Quintile, 2005

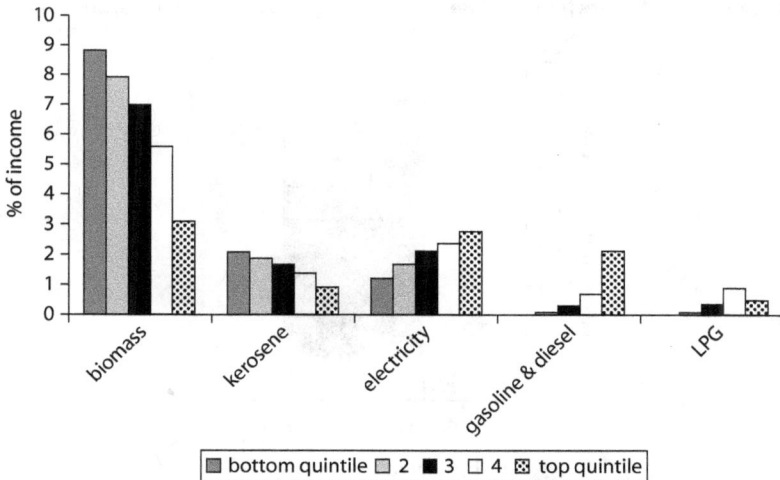

(continued next page)

Figure 3A.12   *(continued)*

b. Urban

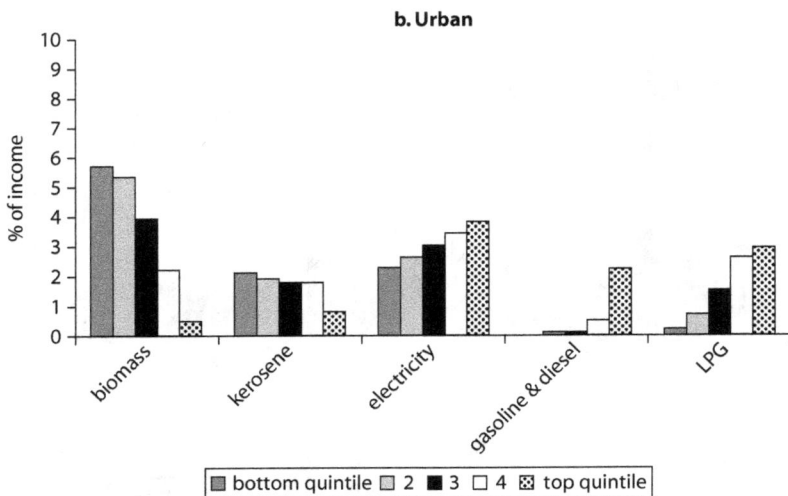

Legend: bottom quintile □ 2 ■ 3 □ 4 ⊠ top quintile

Source: NSSO 2007.
Note: LPG = liquefied petroleum gas.

Figure 3A.13   Welfare Impact of Removing Energy Subsidies in India, 2005

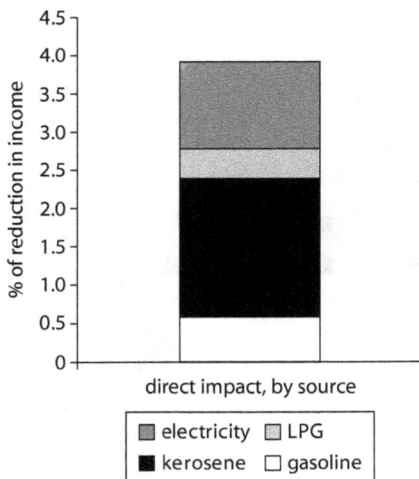

direct impact, by source

Legend: ■ electricity □ LPG ■ kerosene □ gasoline

Source: Del Granado, Coady, and Gillingham 2010.
Note: LPG = liquefied petroleum gas.

References

Bacon, Robert, and Masami Kojima. 2006. "Coping with Higher Oil Prices." Report 323/06, Energy Sector Management Assistance Program and World Bank, Washington, DC.

Del Granado, J., D. Coady, and R. Gillingham. 2010. "The Unequal Benefits of Fuel Subsidies: A Review of Evidence for Developing Countries." Working Paper 10/202, International Monetary Fund, Washington, DC.

Gangopadhyay, Shubhashis, Bharat Ramaswami, and Wilima Wadhwa. 2005. "Reducing Subsidies on Household Fuels in India: How Will It Affect the Poor?" *Energy Policy* 33 (18): 2326–36.

Government of India. 2010. "Report of the Expert Group on a Viable and Sustainable System of Pricing Petroleum Products." Report for the Ministry of Petroleum and Natural Gas, Government of India, New Delhi.

IEA (International Energy Agency). n.d. Electricity Price Database. IEA, Paris. http://www.iea.org/stats/index.asp.

NSSO (National Sample Survey Organisation). 2007. "Energy Use by Indian Households 2004/2005." NSS 61st Round Report. Ministry of Statistics and Program Implementation, Government of India, New Delhi.

OECD (Organisation for Economic Co-operation and Development). 2011. "OECD Economic Surveys: India." OECD, Paris.

Pachauri, Shonali, and Jiang Leiwen. 2008. "The Household Energy Transition in India and China." *Energy Policy* 36 (11): 4022–35.

Shenoy, Bhamy V. 2010. "Lessons Learned from Attempts to Reform India's Kerosene Subsidy." Case study paper, Institute for International Sustainable Development, Winnipeg, Manitoba, Canada.

World Bank. 2007. "Social Protection for a Changing India." Two-volume Social Protection Report for the Government of India, South Asia Region, World Bank, Washington, DC.

———. 2010. "Impact of the Global Financial Crisis on Investments in South Asia's Electric Power Infrastructure: The Experience of India, Pakistan, and Bangladesh." Report 56849-SAS, World Bank and Energy Sector Management Assistance Program, Washington, DC.

CHAPTER 4

Jordan

Incentives to Energy Subsidy Reforms

Fossil fuels account for over 90 percent of Jordan's primary energy use. With limited domestic energy resources, Jordan relies heavily on imports. Until 2003, Jordan imported most of its oil from Iraq at discounted rates. The share of oil as primary energy has declined because of the increasing use of natural gas in electricity generation following the start of natural gas imports from Egypt in 2003.

The small share of domestic resources in primary energy supply made Jordan vulnerable to the volatility of international prices of fossil fuels. This vulnerability was enhanced by limited fiscal space and borrowing capacity. Jordan's budget deficit in 2007 reached 7.9 percent of gross domestic product (GDP), while public debt represented 67.5 percent of GDP. The high import dependence also raised energy security risk concerns among policy makers.

The decision to phase out fossil fuel subsidies came after the supply of cheap oil from Iraq came to an end following the United States-led invasion in 2003. The fiscal cost of energy subsidies increased from US$60 million in 2002 to US$711 million in 2005, or 5.6 percent of GDP (see figure 4A.1). To address this major fiscal challenge, the Jordanian government embarked on an ambitious subsidy reform program and decided that the subsidies would be phased out between 2005 and 2008, a realistic time-line that the government closely met. Jordan also took several measures to curb demand growth (fueled by rising GDP per capita) by adopting significant demand and supply measures, which are embedded in the updated National Energy Strategy adopted in December 2007.

Reform Efforts

Fossil Fuels

Unlike most of the countries in the Middle East, petroleum pricing in Jordan is cost-reflective, and the prices have been adjusted to reflect international benchmarks, although heavy fuel oils for electricity genera-tion, diesel, and liquefied petroleum gas (LPG) have been subsidized. These subsidies were phased out in 2008, with the notable exception of LPG because LPG is used by low-income households for cooking. Kerosene was used by households mainly for heating but has been partly replaced by electricity.

Since March 2008, retail petroleum prices have been adjusted monthly based on a formula for an international benchmark netback value. Fossil fuel prices reached a peak in August 2008 and fell gradu-ally since then, with the exception of gasoline whose prices continued to increase, reaching a level now higher than the one in the United States (see figure 4A.2). Fossil fuel prices have been frozen since late 2010 to shield consumers from the volatility in oil prices. Phasing out energy subsidies had reduced the subsidy bill from 5.6 percent of GDP in 2005 to 0.4 percent in 2010 (see figure 4A.1).

A different trend in the consumption of diesel and gasoline fuels in the road sector is observed. Whereas diesel road consumption declined from a peak in 2005 to pre-2003 levels in 2008 (figure 4A.3), gasoline road consumption continued to increase from 2005, most likely because of the increase in the number of gasoline-powered vehicles in Jordan (figure 4A.4).

Electricity

Electricity tariffs are largely cost-reflective, with some cross-subsidies embedded in the tariff structure that benefit households that consume low levels of electricity. Currently, electricity is available for almost all households while LPG is the main fuel used for cooking and water heat-ing (NEPCO 2009). The power industry has been unbundled into gen-eration, transmission, and distribution segments. The largest generation company, the Central Electricity Generating Company, and all three distribution companies have been privatized.

The increase in nominal tariff rates in March 2008 (figure 4A.5) was progressive (Besant-Jones 2009). The tariff rates rose only slightly for the lowest consumption block; by about 20 percent and 37 percent for the second and third blocks, respectively; and by 38 percent for the highest block (see figure 4A.7).

Electricity tariffs remained mostly cost-reflective (with some cross-subsidies) from 2008 to 2010. However, the levels of subsidy embedded in the tariffs increased because of (a) the continued interruption of Egyptian gas supplies to the power generation facilities in Jordan during 2011; and (b) the use instead of diesel and heavy fuel oil, with the implied increase in generation keeping the electricity tariffs unchanged. In July 2011, the government allowed an increase in the domestic electricity tariffs for the consumption block above 750 kilowatt-hours (kWh).

It is too early to assess the impact of the price change on power consumption, but the existing evidence for 2008 shows that power consumption was still increasing, fueled by higher GDP per capita (see figure 4A.6)

Poverty Alleviation Measures

Palliative measures, largely regarded as successful in dealing with price increases, have included the following:

- The minimum wage was increased, and low-paid government employees received higher wage increases than other employees.
- A one-time bonus was given to low-income government employees and pensioners.
- An electricity lifeline tariff for those using less than 160 kWh per month was maintained with the help of cross-subsidization.
- Cash transfers were provided to other low-income households whose heads were nongovernmental workers or pensioners.
- Tax exemptions were implemented that aimed at low-income groups by targeting 13 basic foodstuffs.
- Government funding was increased to the National Aid Fund (NAF) as part of a program to improve the design and implementation of this national safety net program with the World Bank's assistance (Coady et al. 2010).

- The impact on the nonpoor was addressed by removing the government sales tax on nontourist restaurants and temporarily removing it for retailers with annual turnover below US$1.4 million, for taxis, and for public transport.
- Measures aimed at fuel substitution and energy efficiency were implemented along with subsidy reform. Taxis were permitted to increase their prices, and the cost of public transport also rose.

Evidence from Household Surveys

The availability of two household surveys before and after 2008 provides the opportunity to examine the direct impact of the March 2008 tariff increase on household consumption (Besant-Jones 2009). Households switched from kerosene and diesel to electricity when the relative price of fuels increased but the electricity price was left unchanged. Besant-Jones (2009) reports evidence that consumption across low-income households is more elastic than across high-income households and is also sensitive to changes in the average price of electricity.

A comparison of the 2006 and 2008 data shows that the distribution of electricity consumption by decile became progressive in 2008 after the tariff reforms, with expenditures by high-income households becoming higher than those of low-income households (see figures 4A.8 and 4A.9). In other words, the rising block rate for tariffs is an effective means of providing subsidies to low-income consumers. Consumption of LPG does not vary with annual seasons, increases moderately with income, and is characterized by high price elasticity. No definite conclusions can be reached for diesel and kerosene. The switch from kerosene to electricity can be attributed either to the price increase of kerosene relative to electricity or simply to the absence of space heating during the months covered by the 2008 survey.

Fossil Fuels

Del Granado, Coady, and Gillingham (2010) consider the direct and indirect impacts of a US$0.25 per liter increase in fuel prices in the case of Jordan. The direct impact of phasing out subsidies—considering only the impact on the consumption of fuels for cooking, heating, lighting, and private transport—is a loss of slightly above 5 percent of real income, mostly due to the reduction in electricity consumption. The indirect impact—through higher prices for other goods and services consumed by households because higher fuel costs are reflected in increased production costs and consumer prices—is slightly higher, just above 6 percent of real income (see figure 4A.10).

Social Safety Nets

Jordan has well-developed, strong social safety net programs compared with many countries in the region. The programs fall into three general categories:

- *Income support* to poor and vulnerable families, implemented by two key institutions: NAF and the Zakāt Fund
- *Social care services* to vulnerable groups such as people with disabilities; children, youth, families, and women in distress; and others
- *Economic empowerment* through skills and asset development, the most important effort of which is the Enhanced Productivity Program hosted by the Ministry of Planning and International Cooperation.

Both public and private providers are involved in safety net program delivery. Total public spending on safety nets is estimated at more than 1 percent of GDP, with about half spent through NAF. The total number of beneficiaries is estimated at about 8–10 percent of the population (World Bank 2008).

Of the programs, the most important reform attempts have been to streamline the operations of NAF, which the World Bank has also been involved in. NAF was established in 1986 to provide cash social assistance to the poor and played a crucial role as a mechanism to reduce negative welfare effects on the poor during the fuel subsidy phaseout. There is still some room for improvement to reduce leakages to the nonpoor (World Bank 2008).

Key Lessons Learned

Jordan is the one of the few countries in the Middle East and North Africa region to have succeeded in implementing fuel subsidy reforms. Price regulation was replaced with automatic monthly adjustments of the domestic price to reflect changes in international prices—a strategy that has generated significant savings for the government. While the transition has been relatively smooth and peaceful, the higher prices have not passed without any grievances. Because the pricing of petroleum products is now based on a monthly adjustment of the domestic prices to reflect the international price, this has caused some dissatisfaction for gasoline station owners who complained about being asked to sell the products bought at the higher prices in the international market for lower prices in the national market. From the consumer perspective, gasoline station owners refuse to sell fuel in the days prior to the expected increase in prices.

Although Jordan had no other options but to reform its fuel subsidies since the price of these commodities (on which it so heavily depends) changed so abruptly because of the Iraq war, there are merits for Jordan's undertaking and implementation of reforms in a timely fashion. Part of the reason why Jordanian reform was successful is its previous experience with subsidy withdrawals such as its food subsidy reform of the 1990s, for which the government used its cash transfer program.

Another key element in the reform was the large public communication campaign undertaken to inform the people and prevent protests. A wide-ranging compensation package was introduced to prevent increases in poverty and to secure the consent of the nonpoor. In absolute terms, energy subsidies were highly regressive, but the poor spent a higher proportion of their income on fuel. Reforms were also successful because they were coordinated with various stakeholders and preceded by consultations with parliament, the local nongovernmental organizations, the business community, and labor representatives. The political will and determination to phase out subsidies and not backtrack on reform was there, as the emphasis was placed on the regressive nature of the subsidy and the leakage to higher-income groups.

Annex 4.1 Jordan Case Study Figures

INCOME LEVEL: Lower-middle income
REGION: Middle East and North Africa
ENERGY NET IMPORTER/EXPORTER: Net importer
SUBSIDIES: Oil
PHASING OUT SUBSIDIES: Successful

Fiscal Burden of Energy Subsidy in Jordan

Figure 4A.1   Explicit Budgetary Energy Subsidies in Jordan, 2002–10

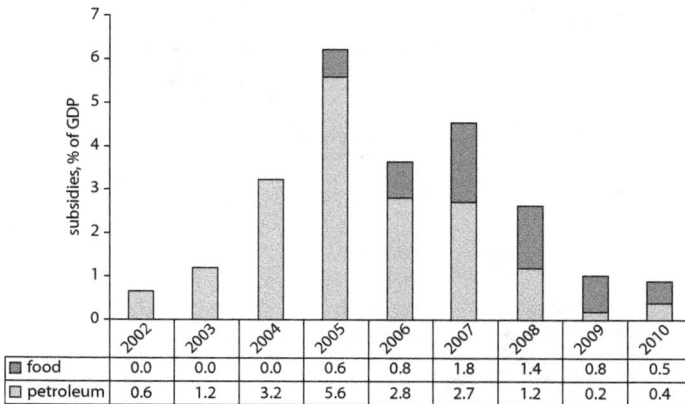

	2002	2003	2004	2005	2006	2007	2008	2009	2010
food	0.0	0.0	0.0	0.6	0.8	1.8	1.4	0.8	0.5
petroleum	0.6	1.2	3.2	5.6	2.8	2.7	1.2	0.2	0.4

Sources: Ministry of Finance; IMF 2009; and other IMF staff reports.

Fuel Prices and Road Sector Consumption in Jordan

Figure 4A.2   Domestic Retail Fuel Prices in Jordan, 2002–10

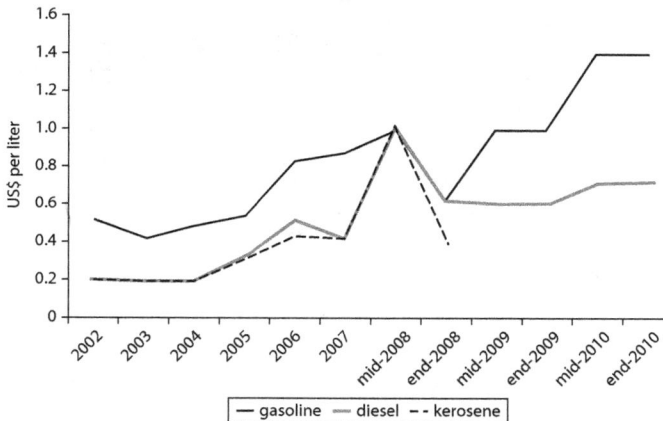

— gasoline — diesel -- kerosene

Sources: Based on data from GIZ n.d.; IMF 2010; and additional data from individual country information.

Figure 4A.3   Road Sector Diesel Consumption in Jordan, 1998–2008

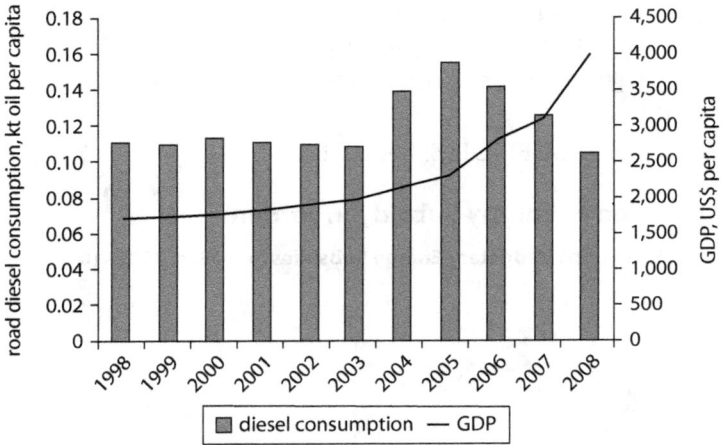

Source: World Bank, World Development Indicators.

Figure 4A.4   Road Sector Gasoline Consumption in Jordan, 1998–2008

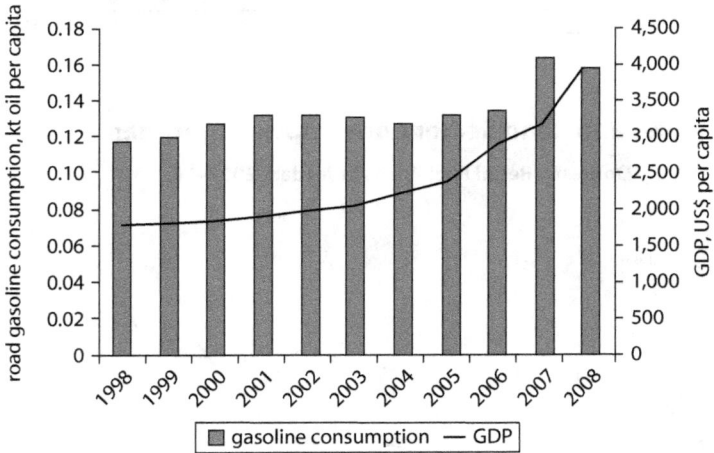

Source: World Bank, World Development Indicators.

Electricity Price and Power Consumption in Jordan

Figure 4A.5   Electricity Price in Jordan, 2002–10

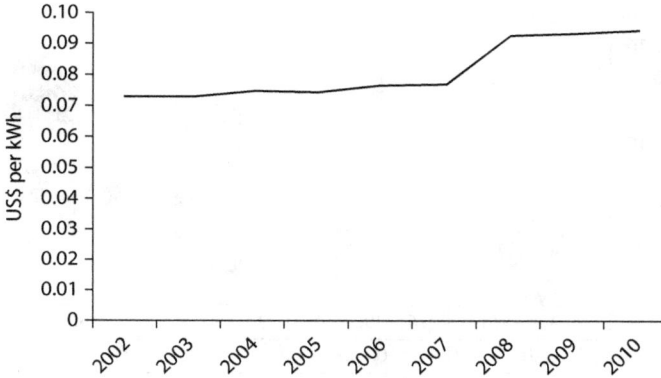

Source: Electricity Regulatory Commission annual reports.
Note: kWh = kilowatt-hour.

Figure 4A.6   Power Consumption Per Capita in Jordan, 1998–2008

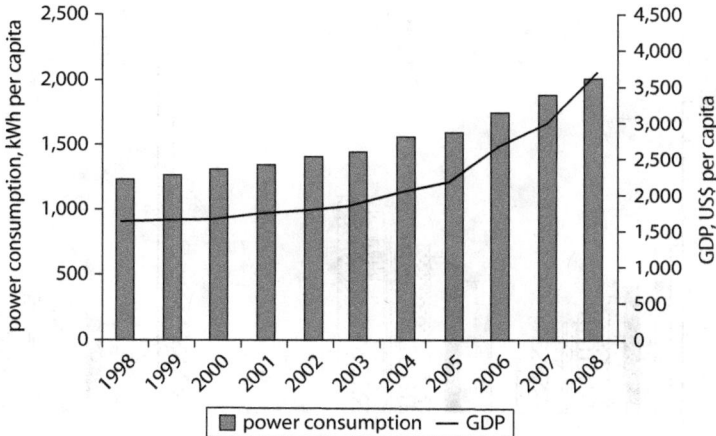

Source: World Bank, World Development Indicators.
Note: kWh = kilowatt-hour.

Poverty Impact Evidence from Household Surveys in Jordan

Figure 4A.7   Electricity Block Tariffs in Jordan, 2008

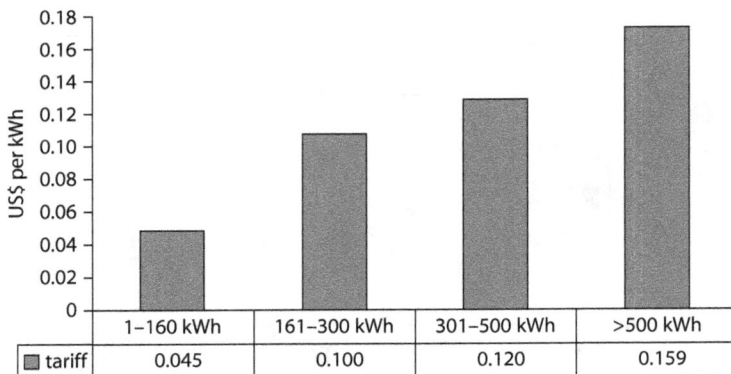

	1–160 kWh	161–300 kWh	301–500 kWh	>500 kWh
tariff	0.045	0.100	0.120	0.159

Source: Tariff schedule approved by the Electricity Regulatory Commission, March 2008.
Note: kWh = kilowatt-hour.

Figure 4A.8   Energy Expenditure in Jordan, by Income Decile, 2006

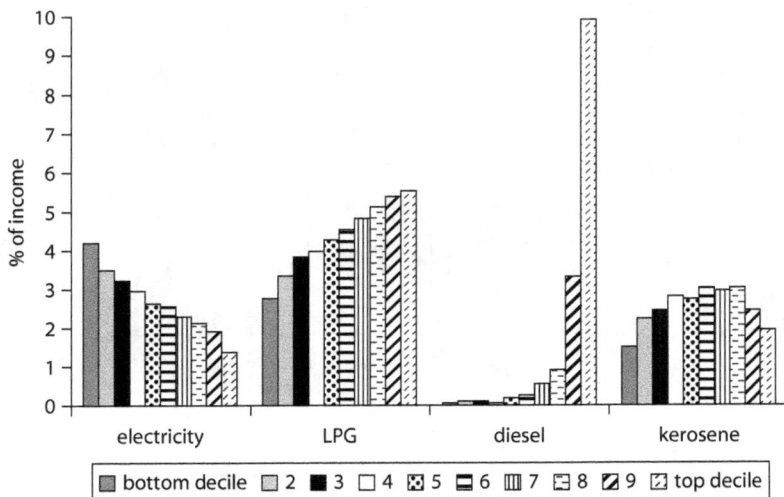

Source: Besant-Jones 2009 from July–December 2006 household survey data.
Note: LPG = liquefied petroleum gas.

Figure 4A.9 Energy Expenditure in Jordan, by Income Decile, 2008

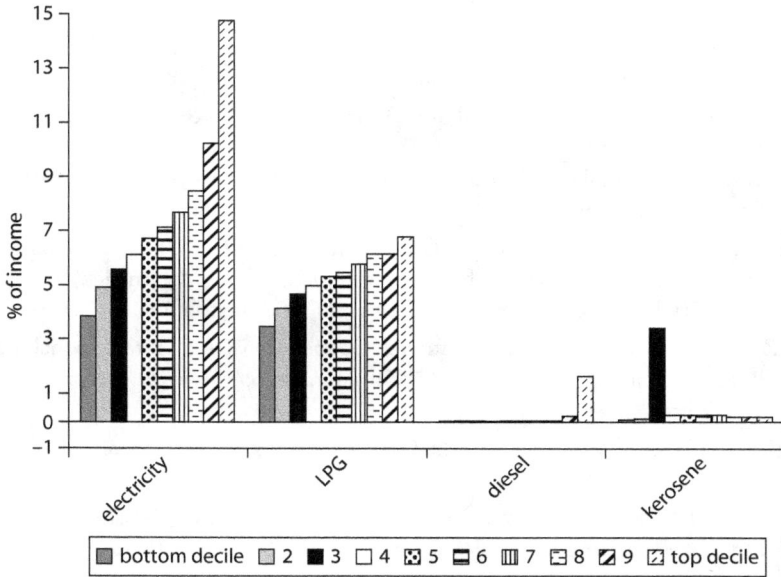

Source: Besant-Jones 2009 from March–May 2008 household survey data.
Note: LPG = liquefied petroleum gas.

Figure 4A.10 Welfare Impact of Removing Fossil Fuel Subsidies in Jordan, 2002

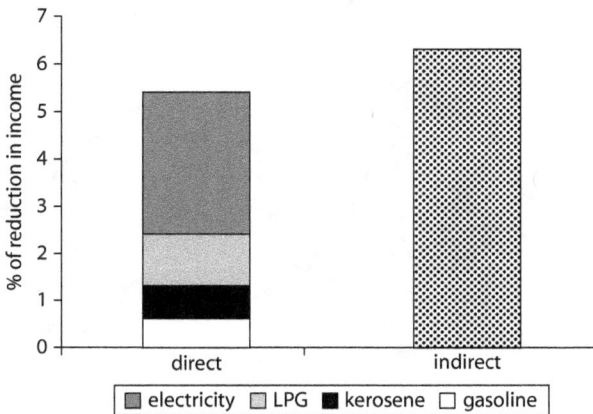

Source: Del Granado, Coady, and Gillingham 2010, based on 2002 household survey data.
Note: LPG = liquefied petroleum gas. Indirect impacts include higher prices for non-energy goods and services consumed by households to the extent that increased production costs and consumer prices reflect higher fuel costs.

References

Besant-Jones, John. 2009. "Jordan—Price Shocks and Subsidy Reform: Poverty and Fiscal Impact Study. Household Energy Sector Analysis." Unpublished working paper, World Bank, Washington, DC.

Coady, David, Robert Gillingham, Rolando Ossowski, John Piotrowski, Shamsuddin Tareq, and Justin Tyson. 2010. "Petroleum Fuel Subsidies: Costly, Inequitable, and Rising." Staff Position Note, International Monetary Fund, Washington, DC.

Del Granado, J., D. Coady, and R. Gillingham. 2010. "The Unequal Benefits of Fuel Subsidies: A Review of Evidence for Developing Countries." Working Paper 10/202, International Monetary Fund, Washington, DC.

GIZ (German Agency for International Cooperation). n.d. International Fuel Prices database. GIZ (formerly GTZ), Bonn. http://www.gtz.de/en/themen/29957.htm.

IMF (International Monetary Fund). 2009. "Jordan: Article IV Consultation." IMF, Washington, DC.

————. 2010. Retail domestic fuel prices data sheet, IMF, Washington, DC. http://www.imf.org/external/pubs/ft/spn/2010/data/spn1005.csv.

NEPCO (National Electricity Power Company). 2009. Annual Reports. Amman, Jordan.

World Bank. 2008. "Social Protection Enhancement in Jordan." Project Appraisal Document, Human Development Sector, Middle East and North Africa Region, World Bank, Washington, DC.

CHAPTER 5

Moldova

Incentives to Energy Subsidy Reforms

Despite a relatively low level of energy consumption per capita, Moldova's economy has a high energy intensity, using twice as much energy per unit of gross domestic product (GDP, at purchasing power parity rate) than Romania and three times more than the European Union (EU) average (World Bank 2009). Moldova is heavily dependent on energy imports, with more than 95 percent of its energy imported from the Russian Federation and Ukraine. The increase in Russian and Ukrainian gas and oil prices in the mid-1990s accumulated about US$300 million in debt by the state energy company, Moldenergo, which kept residential tariffs low until 1998. Moldova was facing an energy crisis with supply shortages and power interruptions (Lampietti, Banerjee, and Branczik 2007). This was one of the main pressures to reform the Moldovan power sector.

Major reforms were undertaken entailing unbundling, privatization, and establishment of a sound regulatory framework. The independent National Energy Regulatory Agency (ANRE) was established in 1998 for the purpose of regulating the electricity, natural gas, and district heating subsectors. ANRE adopted cost-recovery tariffs using a rate-of-return methodology. However, until recently, the municipalities remained in charge of setting the tariffs for district heating and remained heavily subsidized.

Hidden costs coming from underpricing, lack of collection, and unaccounted losses were substantially reduced, dropping from about 11 percent of GDP in 2000 to below 3 percent of GDP in 2003 (see figure 5A.1). The largest component of hidden costs comes from unaccounted losses, which were also reduced substantially from more than 7 percent of GDP in 2000 to 2.7 percent of GDP in 2003. Hidden costs due to collection failures declined from more than 2 percent to insignificant levels. Since 2002, electricity prices have been above cost recovery, removing this component of the hidden cost (Ebinger 2006). Natural gas hidden costs remained constant over time, mainly because of collection failures.

The accumulation of arrears is caused by underpayment of heating bills by the Chişinău municipality, Moldova's capital and largest city. The increase in energy import prices has contributed to the accumulation of new debts, mainly in the district heating in Chişinău, where current heating tariffs cover only about 70 percent of the cost. Despite the price increase, natural gas will remain the most competitive fuel for heating and electricity generation, given the existing energy infrastructure (World Bank 2009). The municipality has committed to a schedule of payments to eliminate both these and older heating-related arrears (IMF 2011). The Ministry of Economy, the Chişinău municipality, and key companies from the thermal energy sector, in consultation with the World Bank and the International Monetary Fund, are working on a restructuring plan for this sector.

Reform Efforts

Fossil Fuels

Prices of diesel and gasoline were substantially increased to reflect the international benchmark. As of the end of 2010, the price of diesel was above the U.S. price level, even if still lower than the price levels of Luxembourg and Romania (the two countries characterized by the lowest prices of diesel and gasoline, respectively, in the EU). A different trend in the consumption of diesel and gasoline fuels in the road sector is observed. Whereas diesel road consumption continued to increase from 2003, fueled by an increase in GDP per capita (see figure 5A.3), gasoline road consumption remained rather stable over time, with a sharp decline from a peak in 2003 to pre-2003 levels through 2008 (see figure 5A.4).

The largest energy sector reforms in Moldova took place between 1997 and 2003. The objective of the reforms was the full commercialization of the energy supply, accompanied by social policies to protect the most vulnerable groups. The World Bank played a crucial role in providing funding and advice for the reforms. Below are the major reforms that took place in this period.

Electricity

- To develop a market-based legal framework in 1998, the regulatory, policy, and ownership functions were separated, allowing private ownership in the energy sector.

- An independent energy regulatory agency (ANRE) was established in 1998 but did not start operating until 2000.

- To achieve restructuring and corporatization, the country's vertically integrated electricity monopoly was unbundled into five distribution companies; four generation companies (three combined heat-and-power plants and a small hydropower plant); and Moldelectrica, a transmission and dispatch company.

- To restructure debt, historic debts of the electricity sector accumulated by distribution companies in the early and mid-1990s were transferred and consolidated on the books of Moldtranselectro, which became a debt holding company.

- The level and the structure of tariffs have been adjusted. Residential electricity tariffs in 1996 were between US$0.012 and US$0.016 per kilowatt-hour (kWh) (average tariffs were below US$0.03 per kWh). This implied significant cross-subsidies to residential consumers for energy consumption in 1997. Tariffs were increased and equalized for all consumers to about US$0.05 per kWh, the level that was more or less maintained from 1997 until 2006. Since then, tariffs have more than doubled, reaching a level above US$0.12 per kWh. The impact of these price increases on power consumption was to decrease consumption by more than 15 percent (see figure 5A.6).

- The existence of a breakaway region of Transdniestria has resulted in a dual energy system in Moldova (EBRD 2004a). Electricity tariffs of Transdniestria are not controlled by ANRE but by its local autonomous administration. Moldova has a secondary energy transportation company, Dnestrenergo, which serves Transdniestria. Transdniestrian authorities also control Moldova's largest plant, Moldavskaya GRES on the left bank of the River Dniester. The plant has operated since 1964 and consists of 12 power generation units and an installed capacity of 2,520 megawatts, not all of which is in use currently. Moldavskaya GRES was privatized and is now owned by the Russian company Inter Rao UES, which seeks to further increase electricity exports.

- Privatization of electricity distribution companies increased infrastructure investment. In February 2000, three of five regional electricity distribution companies (REDs), covering about 70 percent of the market—RED Chişinău (serving the capital region), RED Centru (serving central Moldova), and RED Sud (serving southern Moldova)—were sold for US$26 million in an open tender to the Spanish utility Unión Fenosa, which as part of the deal committed to invest US$56 million in infrastructure rehabilitation over five years (Lampietti, Banerjee, and Branczik 2007).

Privatization results were initially mixed. Tariffs were set according to the terms of the privatization contract with Unión Fenosa in order to guarantee a 23 percent return on investment and reduce regulatory risk for the investor (EBRD 2004b). However, conflicts quickly emerged because tariff adjustments of 2002 and 2003 were delayed and were lower than expected by Unión Fenosa. In addition, a challenge to the legality of the privatization of the distribution companies was not resolved until October 2003. The government still controls certain aspects of ANRE activities, despite the de jure independence of ANRE, through its approval of the annual budget and appointment and dismissal of ANRE directors (EBRD 2004a). Despite the presence of Unión Fenosa, foreign investors are not encouraged to invest in Moldova's energy sector because of government interference and an unstable political system.

Natural Gas

- Moldova's fiscal situation was strengthened through the divestiture of the gas industry. The majority of shares in Moldovagaz, the country's monopoly gas supplier, were sold to Russia's Gazprom in exchange for a portion of the debt.

- With the 2008 increase in Russian natural gas prices—from US$192 to US$233 per 1,000 cubic meters—ANRE decided to increase tariffs by an average of 20.9 percent for electric energy and 29.2 percent for heating. The electricity tariffs for both the Unión Fenosa distribution networks and the state-owned distribution networks increased by about the same amount. Moldovagaz has requested that ANRE increase natural gas tariffs to prevent accumulation of financial losses and increasing debts of the company to Gazprom.

Poverty Alleviation Measures

Evidence from Household Surveys

Almost all households are connected to the electricity grid, while connection rates for central heating and central gas and for the consumption of liquefied petroleum gas (LPG) vary (see figure 5A.7). While more than 70 percent of households in large cities (Chişinău and Balti) and small towns are connected to central gas services, only 11 percent of rural houses are connected. Most rural households consume gas cylinders or LPG, while few households in large cities consume them. About a quarter of all households and over three-quarters of households in small towns and rural areas consume LPG. Central heating, on the other hand, is mainly an urban consumption item (Baclajanschi et al. 2007).

Electricity expenditures are more burdensome for poorer households and account for a larger share of household expenditure than all other energy products combined (see figure 5A.8, panel a). Spending on electricity is somewhat regressive: the bottom quintile spends a budget share of 5.4 percent on electricity while the richest quintile spends only 3.6 percent. In contrast, the richest quintile spends a larger budget share on central heating (1.8 percent) than the lowest quintile (0.1 percent). The expenditure patterns for central gas and LPG, on the other hand, are neither clearly regressive nor progressive.

Focusing only on households that are connected and have positive expenditures on gas and energy products reveals that electricity expenditures are more burdensome for poorer households across all locations (see figure 5A.8, panel b). Also, expenditure on LPG is consistently more burdensome for the poor, accounting for at least 10 percent of the budgets of the poorest quintiles. Central heating, on the other hand, is mostly a consumption item in large cities and is neither clearly progressive nor regressive (see figures 5A.7 and 5A.8).

Social Safety Nets

Moldova's strategy for mitigating the impact of tariff increases was implemented through the Nominative Targeted Compensation (NTC) system, the Moldovan government's primary instrument for delivery of social benefit assistance. The system is not well targeted at the poor (Lampietti, Banerjee, and Branczik 2007). Instead of being means tested, it is a system of categorical privileges: certain groups of people receive the NTC, which helps cover the cost of electricity, gas, district heating, hot water, cold water, coal, and firewood. The proportion of households in the lowest quintile receiving the NTC was only slightly higher than for the highest quintile: 16 percent versus 14 percent. Moreover, the lowest quintile of households received the smallest share of NTC resources, while the highest quintile received the largest (Lampietti, Banerjee, and Branczik 2007).

The government is expected to extend heating assistance to low-income beneficiaries representing over 17 percent of the population. Moreover, enrollment in the targeted social assistance scheme was to expand from 38 percent of eligible households at the end of 2010 to 50 percent in 2011 and 65 percent in 2012 (IMF 2011).

Key Lessons Learned

The overall results of the restructuring of the Moldovan energy sector were positive, with payment collections increasing, especially for electricity in both privatized and state-owned electricity distribution companies (World Bank 2003). Privatization of the distribution had a positive impact on both the service quality and the government budget. The poor benefited more than the nonpoor from the reforms, having increased their consumption more than the nonpoor despite rising costs (Lampietti, Banerjee, and Branczik 2007). During the reforms, between 1998 and 2003, energy efficiency improved as high-volume consumers reduced consumption following the price increases. Tariff increases such as the one taking place more recently may have significant welfare effects for poor households with constant incomes, which underscores the importance of improved social safety nets to address the present issues.

The Moldovan energy sector reform seems to have been triggered by the increase in imported gas prices from Russia, which made the vertically integrated national electricity company unable to operate at the cost-reflective level, resulting in further accumulation of unsustainable arrears and poor electricity supply service. Removing preferential gas pricing can thus exert external pressure to restructure the energy sector by unbundling and partially privatizing it, in due course helping to phase out energy subsidies.

Annex 5.1 Moldova Case Study Figures

INCOME LEVEL: Lower-middle income
REGION: Europe and Central Asia
ENERGY NET IMPORTER/EXPORTER: Net importer
SUBSIDIES: Electricity, gas
PHASING OUT SUBSIDIES: Mostly successful

Fiscal Burden of Energy Subsidy in Moldova

Figure 5A.1 Implicit Subsidies of the Power Sector in Moldova, 2000–03

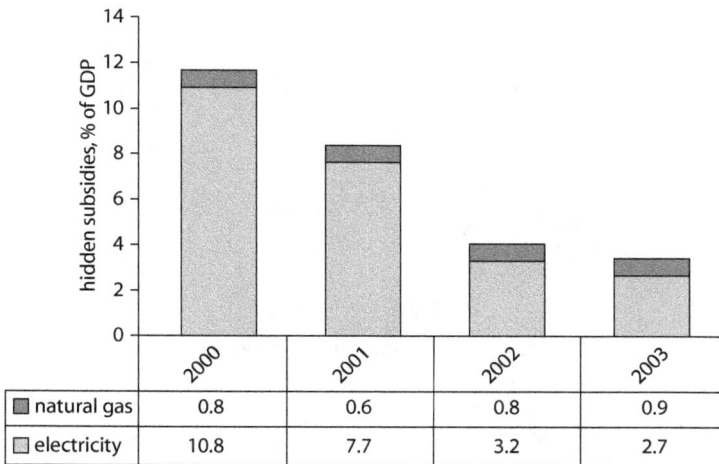

	2000	2001	2002	2003
natural gas	0.8	0.6	0.8	0.9
electricity	10.8	7.7	3.2	2.7

Source: Ebinger 2006.
Note: Implicit subsidies (or hidden costs) are defined as the difference between actual receipts and the revenue that the energy company (for example, a utility involved in the distribution of electricity and natural gas) would receive were it to be in operation with cost-recovery tariffs based on efficient operation with normal losses and with full bill collection.

Fuel Prices and Road Sector Consumption in Moldova

Figure 5A.2   Domestic Retail Fuel Prices in Moldova, 2002–10

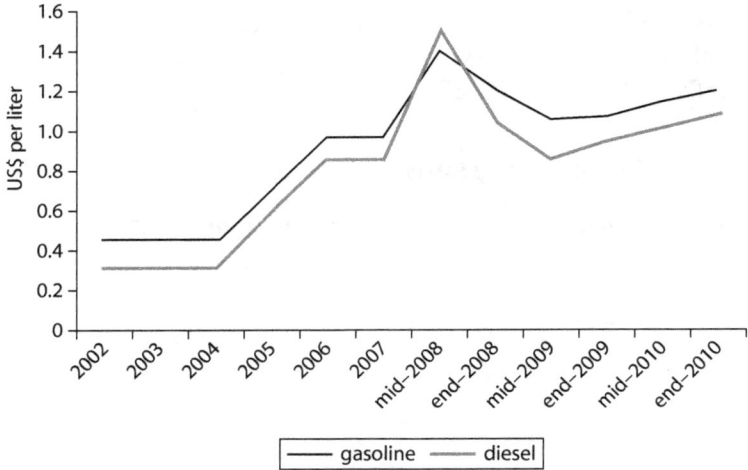

Source: ANRE.

Figure 5A.3   Road Sector Diesel Consumption in Moldova, 1998–2008

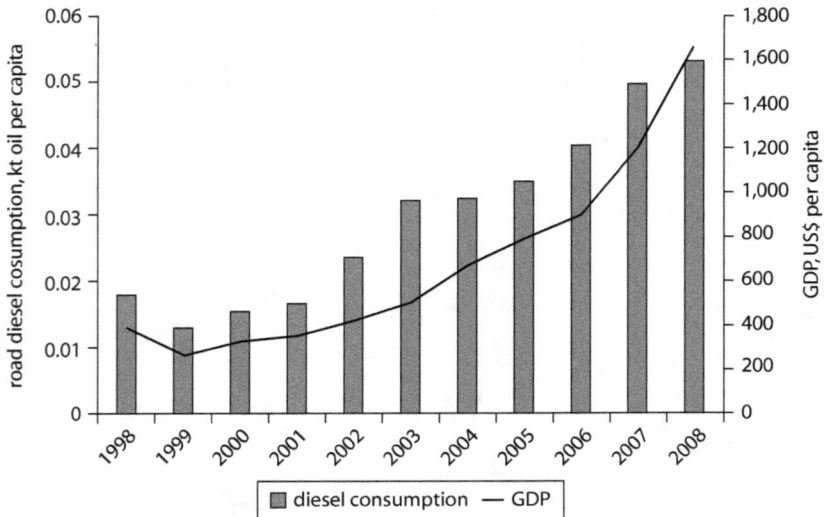

Source: World Bank, World Development Indicators.

Figure 5A.4   Road Sector Gasoline Consumption in Moldova, 1998–2008

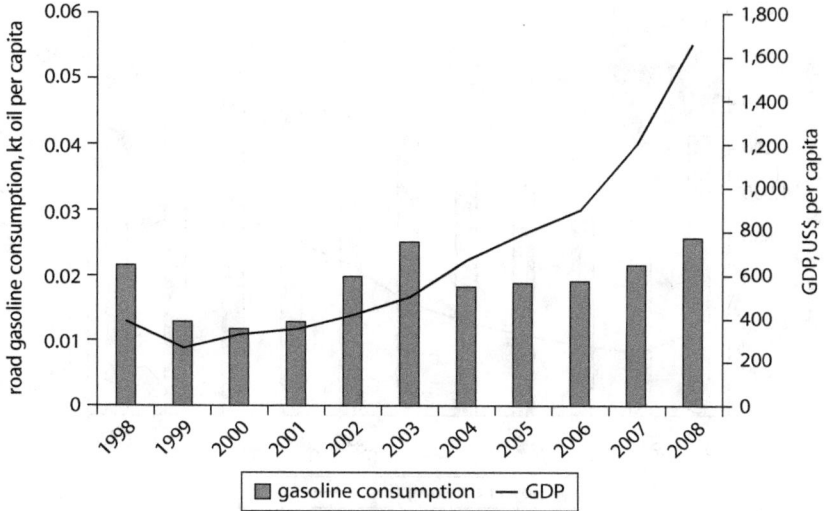

Source: World Bank, World Development Indicators.

Electricity Price and Power Consumption in Moldova

Figure 5A.5   Electricity Price in Moldova, 1999–2010

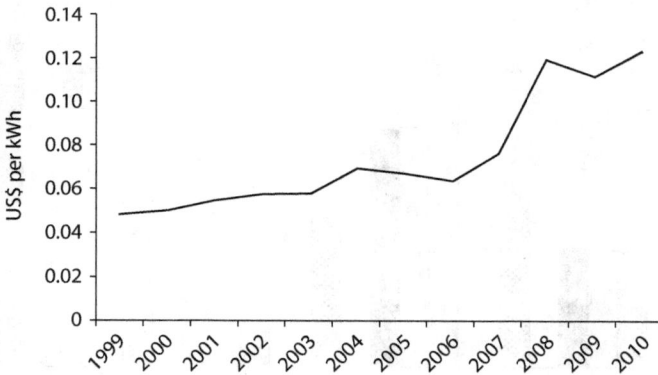

Source: ANRE.
Note: kWh = kilowatt-hour.

Figure 5A.6   Power Consumption Per Capita in Moldova, 1998–2008

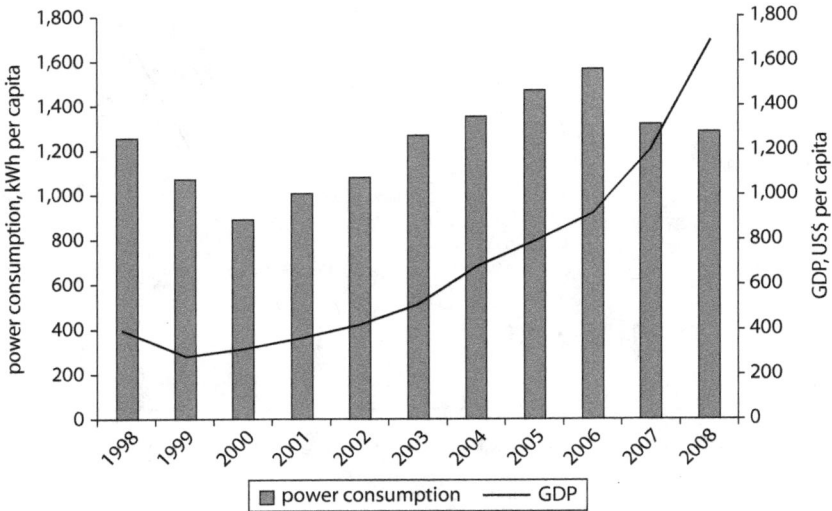

Source: World Bank, World Development Indicators.
Note: kWh = kilowatt-hour.

Poverty Impact Evidence from Household Surveys in Moldova

Figure 5A.7   Households Connected to Energy Sources in Moldova, by Income Quintile, 2004

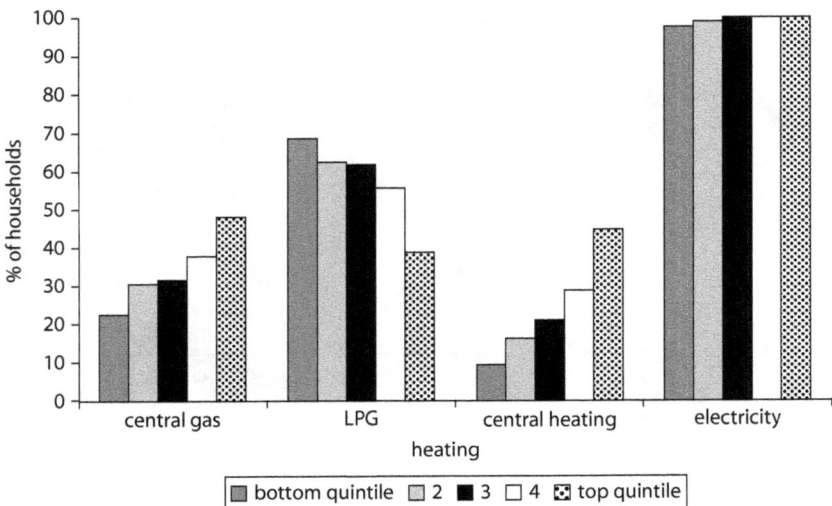

Source: Baclajanschi et al. 2007, from 2004 household budget survey.
Note: LPG = liquefied petroleum gas.

Figure 5A.8   Household Energy Expenditure in Moldova, by Income Quintile, 2004

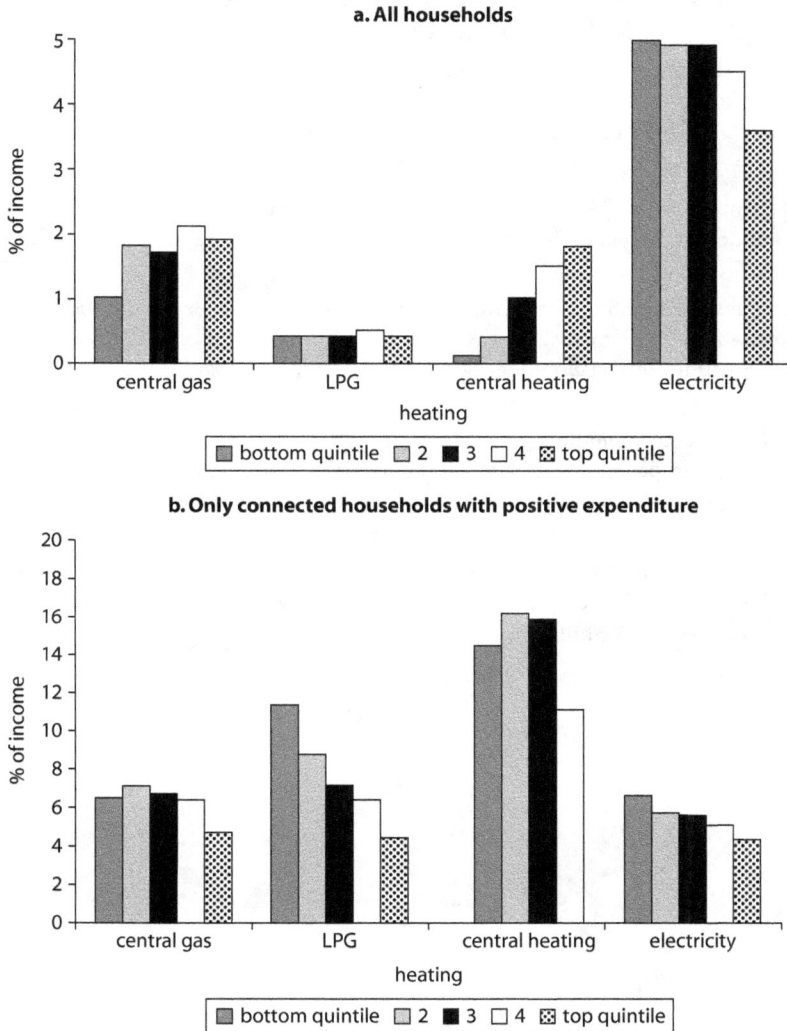

a. All households

b. Only connected households with positive expenditure

Source: Baclajanschi et al. 2007, from 2004 household budget survey.
Note: LPG = liquefied petroleum gas.

References

Baclajanschi, I., L. Bouton, H. Mori, D. Ostojic, T. Pushak, and E. R. Tiongson. 2007. "Rising Energy Prices in Moldova: Macroeconomic and Distributional Impact." *Problems of Economic Transition* 49 (10): 5-40.

Ebinger, Jane. 2006. "Measuring Financial Performance in Infrastructure: An Application to Europe and Central Asia." Policy Research Working Paper 3992, World Bank, Washington, DC.

EBRD (European Bank for Reconstruction and Development). 2004a. "Power Sector Reform in the ETC: Where Now?" Analysis in the Autumn 2004 issue of *LiT online*, a supplement to *Law in Transition*, EBRD, London.

———. 2004b. *Transition Report 2004: Infrastructure.* London: EBRD.

IMF (International Monetary Fund). 2011. "Republic of Moldova: Third Reviews under the Extended Arrangement." Country Report 11/200, IMF, Washington, DC.

Lampietti, Julian, Sudeshna Ghosh Banerjee, and Amelia Branczik. 2007. *People and Power: Electricity Sector Reforms and the Poor in Europe and Central Asia.* Directions in Development Series. Washington, DC: World Bank.

World Bank. 2003. "Project Appraisal Document for Moldova Energy II Project." Infrastructure and Energy Department; Ukraine, Belarus, and Moldova Country Unit; Europe and Central Asia Region, World Bank, Washington, DC.

———. 2009. "Moldova: Policy Notes for the Government." Notes prepared with contributions from the International Monetary Fund, the U.K. Department for International Development, United Nations Moldova, the Swedish International Development Cooperation Authority, and the European Union, World Bank, Washington, DC.

CHAPTER 6

Morocco

Incentives to Energy Subsidy Reforms

Morocco is almost fully dependent on imports to cover its energy needs. The country imports 96 percent of its total commercial energy supply—of which two-thirds are crude oil and petroleum products, and one-third is coal. The share of power produced from fossil fuels increased from 86 percent in 1998 to 95 percent in 2008. Energy imports have increased steadily, from 92 percent of the country's total commercial energy supply in 1998 to 96 percent in 2008. Both petroleum products and electricity are subsidized by the government. Petroleum products are subsidized through a direct compensation to fuel suppliers that covers the difference between the market price and the fixed price of fuels, and electricity is subsidized through a support to the Office National de l'Electricité (ONE), Morocco's national utility company.

By 2000, the rising prices of imported petroleum products drove the Moroccan government to reestablish government control over domestic prices and to discontinue the indexation of domestic energy prices to international prices. The persistence of high prices has made intervention very expensive and convinced legislators that domestic price policy should be revised. The government decided that domestic prices should be set based on market forces. It was not until September 2006

that the government decided to reduce subsidies for selected petroleum products, including diesel and gasoline, but also to continue to
heavily subsidize liquefied petroleum gas (LPG) (butane).[1]

With the globally rising oil prices, the budgetary envelope for the
petroleum subsidies grew from 0.6 percent of gross domestic product
(GDP) in 2003 to the peak of 3.9 percent of GDP in 2007 (see figure
6A.1). The level of petroleum subsidies declined substantially in 2009 (to
1.5 percent of GDP), mostly due to the decrease in world oil prices.
Petroleum subsidies represented 1.2 percent of GDP in 2010. The overall
level of subsidies is expected to increase in 2011 to about 5.5 percent of
GDP, which is much higher than the estimate of 2.1 percent provided in
the 2011 budget (IMF 2011).

Reform Efforts

Fossil Fuels

Morocco was one of the first countries in the Middle East and North
Africa region to engage in significant reforms in the petroleum sector.
These reforms, which were initiated in the mid-1980s, included the privatization of the national refinery (Société Marocaine des Industries de
Raffinage) and of petroleum product distribution activities. Whereas distribution was organized competitively, refining activities remained
monopolistic, with the benefit of a temporary protection against competing imports until the end of 2008. At present, all prices in the petroleum
product sector are still regulated by the government.

The Caisse de Compensation (CDC) operates under the official remit
of keeping energy prices stable over the long term and adjusts consumer
tariffs only when long-run prices change. The CDC levies taxes upon all
petroleum products and sets prices for consumers. Tax levels vary across
products, which are also used to cross-subsidize prices across consumers.
Lower rates are applied to LPG and diesel. Any CDC deficits are met
using direct budgetary transfers (Kelly 2009).

From 1995 (when the new price indexation regime was adopted) to
1999 (when this regime was suspended), the CDC did not require budgetary support. Suspension of the price indexation in the time of the
increasing world prices of oil resulted in CDC deficits. Subsequent developments in fuel pricing reforms include the following:

- In 2002, a review of the price indexation formulae benefiting the
  importers was permitted to balance the compensation fund. However,

despite rises in prices of domestic petroleum products since 2004, the continuing climb in oil prices has translated into increasing deficits for the fund (World Bank 2007; Kelly 2009).

- Fuel tariffs were not raised—by 3.5 percent—until immediately after the harvesting season in August 2004 (to spare farmers hit by drought). Fuel prices were raised twice in 2005, in May and August. In February 2006, as the fiscal costs of the subsidy became unsustainable, the government again raised fuel prices, leaving the LPG price unchanged.

- In September 2006, the government adjusted the prices of petroleum products to fully reflect international prices. This eliminated the subsidies for kerosene and greatly reduced those for gasoline and diesel, resulting in tariff increases of 9 percent for gasoline, 7 percent for diesel, and 8 percent for kerosene (see figure 6A.2). With the two increases taking place in 2006, the full price indexation has been achieved (World Bank 2007). LPG remained largely exempt from these increases, causing the gap between domestic and international prices to grow still further (World Bank 2007; Kelly 2009).

- In 2007, Morocco returned to indexation for some petroleum products except diesel and LPG. Riots accompanied the price rises in February 2006 and September 2007. In February 2006, the price rise was accompanied by a 7 percent rise in the value added tax levied upon fuels marketed at service stations, which led to a strike by the Moroccan Federation of Fuel Traders. These protests show the sensitivity of policies that phase out subsidies (World Bank 2007; Kelly 2009).

- In 2009, Morocco unveiled a new energy strategy. The principal aims were to ensure energy security, meet the needs of all energy consumers at minimum cost, and address the climate change risk. The country currently relies heavily on imported oil, making it highly vulnerable to fluctuations in international markets. The 2008 commodity price boom provided a catalyst for change (Kelly 2009).[2] Compared with other countries, Morocco's retail prices for fuel have been relatively unaffected by the 2008 peak in world fuel prices (see figure 6A.2). In mid-February 2009, however, the government reduced prices for various fuels by different amounts ranging between 9 percent and 26 percent.

- In 2010, the authorities took early steps to reform the subsidy system. The International Monetary Fund welcomed the authorities' intention to develop a global reform program whereby universal subsidies would be gradually eliminated in parallel with the introduction of new targeted assistance (totaling no more than 2 percent of GDP).

Electricity

The power sector has been successfully opened to private investment for power generation and distribution. The private sector was actively involved in developing the power generation capacity between 1997 and 2003. As a result, the private sector plays a key role in the electricity sector in Morocco, with over 50 percent of the power generating capacity in the hands of independent power producers (contributing to over 70 percent of generation) and 55 percent of electricity distribution done by private operators. Currently, the regulation of the power sector is split between several ministries, including the Ministry of Energy and Mining, the Ministry of Finance and Privatization, the Ministry of Interior for the power distribution state-owned companies, and the Ministry of General and Economic Affairs. The government intends to concentrate the regulatory function in a single regulatory agency (World Bank 2007).

Electricity tariffs increased over time until 2008, reflecting the increasing trend in fuel inputs, but they declined since then. They remain significantly higher than in other countries in the Middle East (including Algeria, the Arab Republic of Egypt, and Tunisia), which has some implications for regional system integration and undermines the effort to accelerate Morocco's integration into the European Union–Maghreb regional power market. The integration would possibly help Morocco to export "green electricity" from wind farms (World Bank 2007).

Poverty Alleviation Measures

Evidence from Household Surveys

Nearly all households (99 percent), both urban and rural, use LPG primarily for cooking and, in rural areas where electricity is not available, also for lighting (see figure 6A.9). Historically, the use of LPG was encouraged by the government to provide incentives to switch away from wood and charcoal, making it the most affordable consumption fuel. Recent surveys indicate a likelihood that the poorest households, especially in the rural areas, would switch from LPG to charcoal, wood, or kerosene if the price of LPG increased to the market level. Two-thirds of the respondents to the surveys stated that they would not change their behavior if the price of the 12 kilogram LPG bottle went up from DH 40 to DH 50. One-third stated that they would change their consumption behavior: households living in rural areas would mostly switch to wood, and urban residents would switch to charcoal since wood collection is not possible in the urban settings. The majority of the respondents (89 percent) did not even know that LPG was subsidized (World Bank 2008a).

Diesel and gasoline subsidies disproportionately benefit higher-income households, which implies high leakages to the nonpoor (IMF 2008; Kelly 2009). Households in the top quintile of the income distribution receive more than 75 percent of diesel subsidies, while the poorest quintile receives less than 1 percent, with gains by the poor from these subsidies related to public transportation likely to be negligible (World Bank 2007).

In case of the LPG subsidies, the richest quintile receives more than 33 percent of government subsidies, while the poorest quintile receives less than 10 percent. At a rural level, the divide between the richest and the poorest quintiles increases even further, with 40 percent of the top quintile receiving LPG subsidies relative to the poorest quintile, which receives only 8.7 percent. For urban households, the richest quintile receives one-third of the subsidies related to LPG, and the poorest quintile receives 10 percent (World Bank 2008a).

Social Safety Nets

Despite a long history of food subsidies, childhood malnutrition remains common, and maternal and infant mortality rates are among the highest in the region. Moreover, many rural communities lack basic infrastructure and access to basic services such as health and education. Morocco has no universal retirement or social protection program, such that the full weight of old age and disability must be borne by the family (Kelly 2009).

In 2003, a new strategy focused on the eradication of slums in order to provide alternative housing for all, although recent advances indicate that the forecasts were overoptimistic. Another aid is provided through programs to promote employment, and regulation of the labor market will be improved through the reform of the National Employment Agency. The National Agency of Social Affairs operates more than 500 social affairs centers across the country, caring for the elderly, female-headed households, and orphans, but it does not disburse sufficient funding (World Bank 2008a).

In 2005 the government of Morocco launched a national program to combat poverty aiming explicitly to correct social distortions brought about by mistargeting of public resources. The general subsidies are provided by the CDC. While costly subsidies still remain the main safety net for Moroccans, new projects will make phasing out of existing subsidies much more feasible (World Bank 2008b, 2009).

Social assistance displays considerable variation in targeting, allocation, and management efficiency. Recent efforts put more emphasis on the problems of rural poverty in order to correct an imbalance that favors social protection for urban, as opposed to rural, populations. Currently, the government is developing new approaches to social safety nets and has tried to pass from price subsidies to income support measures and poverty reduction through services. The growth of the urban population and slums has surpassed the capacity of the social security system (World Bank 2008a).

As economic considerations point toward introducing unconditional cash transfers, a pilot program on education-related conditional cash transfers was launched in rural areas, with transfers beginning in 2009. The program is targeted at the schools in the poorest communities for coverage of 80,000 primary school students (Kelly 2009; World Bank 2009).

Key Lessons Learned

An interesting lesson from the Moroccan case is that although switching to cleaner and more efficient fuels such as LPG is desirable when compared with biomass alternatives, the subsidy on LPG that was used as a transitory measure became permanent.

The government has not yet been able to phase out the LPG subsidy because of underdeveloped social safety net capacities and potential threats of social unrest (World Bank 2007). More recent increases in prices (July 2008) were not accompanied by social unrest like in the past, suggesting that Morocco is making some improvements in the implementation strategy of its subsidy reduction. Improving social safety nets, meeting future electricity demand, and carrying out successful information campaigns will help Morocco to successfully further reduce its subsidy expenditures while shielding vulnerable households.

Morocco is North Africa's largest energy importer in both absolute and relative terms, with imports exceeding 95 percent of the energy supply, mostly in the form of crude oil and coal. Morocco also imports about 18 percent of its total electricity. Morocco is thus highly exposed to international price fluctuations. Development of alternative energy sources such as solar and wind renewable energies, for which Morocco has particular potential, will be important factors in diversifying the energy mix and making electricity more reliable and affordable. The government objective is to generate 20 percent of electricity from renewable energy sources by 2015 (World Bank 2007).

Annex 6.1 Morocco Case Study Figures

INCOME LEVEL: Lower-middle income
REGION: Middle East and North Africa
ENERGY NET IMPORTER/EXPORTER: Net importer
SUBSIDIES: LPG, diesel, gasoline, electricity
PHASING OUT SUBSIDIES: Ongoing

Fiscal Burden of Energy Subsidy in Morocco

Figure 6A.1   Explicit Budgetary Energy Subsidies in Morocco, 2002–10

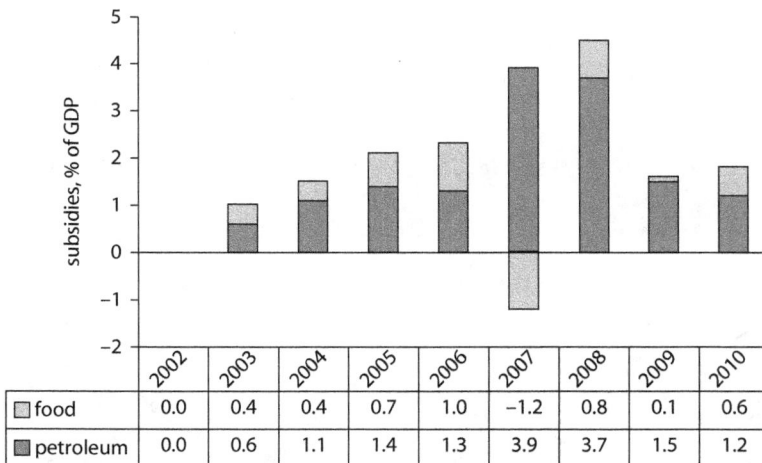

	2002	2003	2004	2005	2006	2007	2008	2009	2010
food	0.0	0.4	0.4	0.7	1.0	−1.2	0.8	0.1	0.6
petroleum	0.0	0.6	1.1	1.4	1.3	3.9	3.7	1.5	1.2

Sources: Based on IMF 2008 and Ministry of Finance information.

Fuel Prices and Road Sector Consumption in Morocco

Figure 6A.2   Domestic Retail Fuel Prices in Morocco, 2002–10

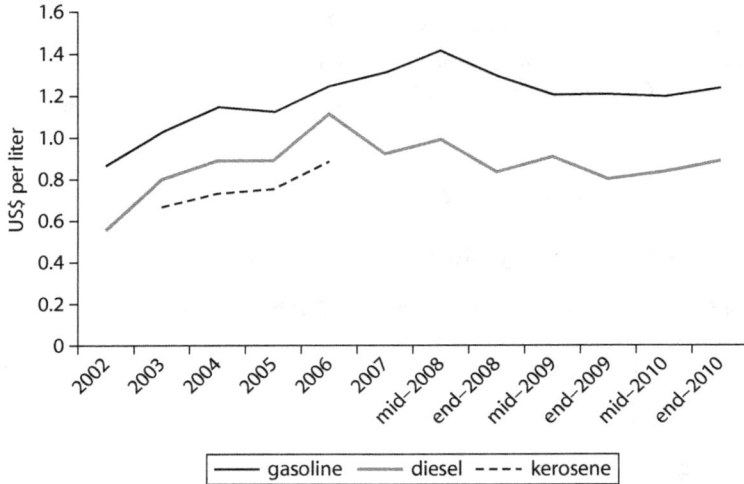

Sources: Elaboration of data from GIZ n.d.; IMF 2010; and additional data from individual country information.

Figure 6A.3   Road Sector Diesel Consumption in Morocco, 1998–2008

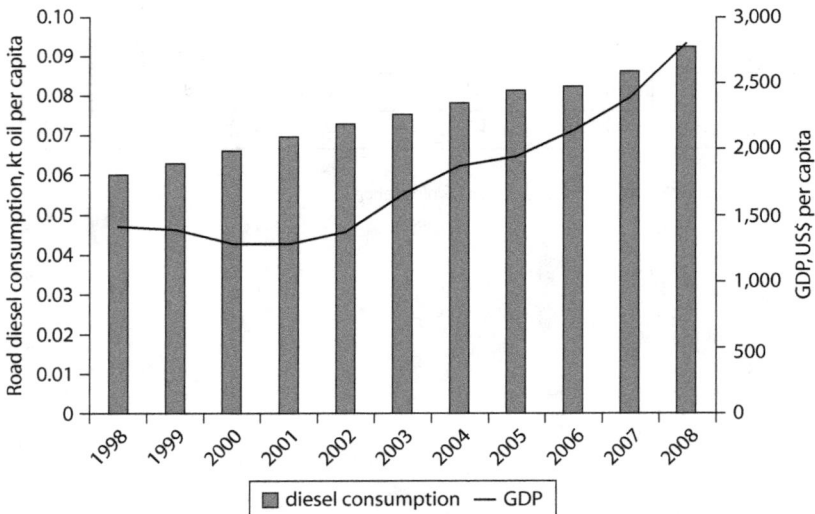

Source: World Bank, World Development Indicators.

Figure 6A.4   Road Sector Gasoline Consumption in Morocco, 1998–2008

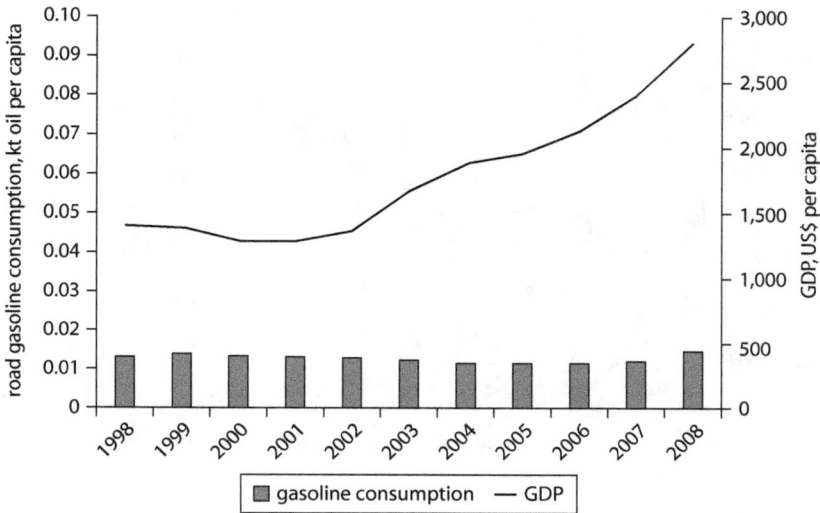

Source: World Bank, World Development Indicators.

Electricity Price and Power Consumption in Morocco

Figure 6A.5   Electricity Price in Morocco, 2002–09

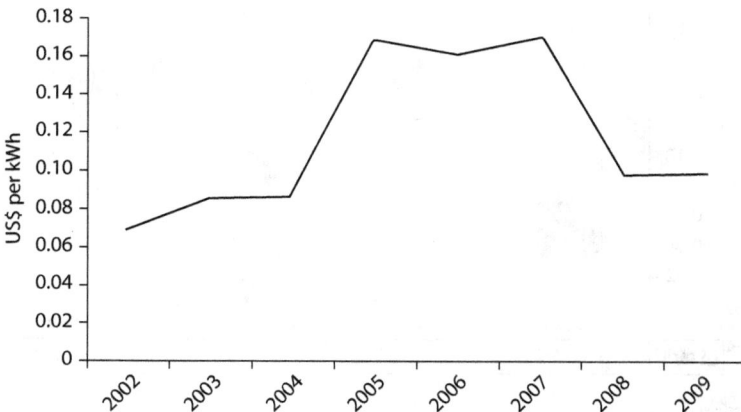

Source: ONE.
Note: kWh = kilowatt-hour.

Figure 6A.6   Power Consumption Per Capita in Morocco, 1998–2008

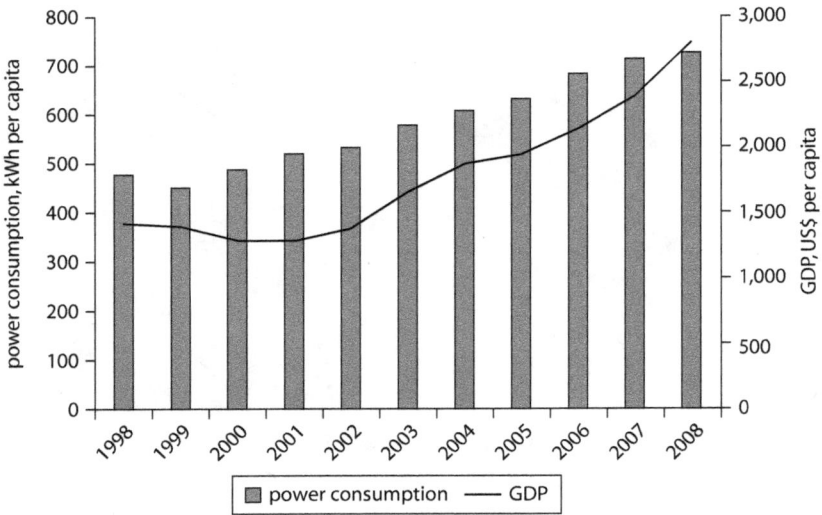

Source: World Bank, World Development Indicators.
Note: kWh = kilowatt-hour.

Poverty Impact Evidence from Household Surveys in Morocco

Figure 6A.7   Electricity Block Tariffs in Morocco, 2010

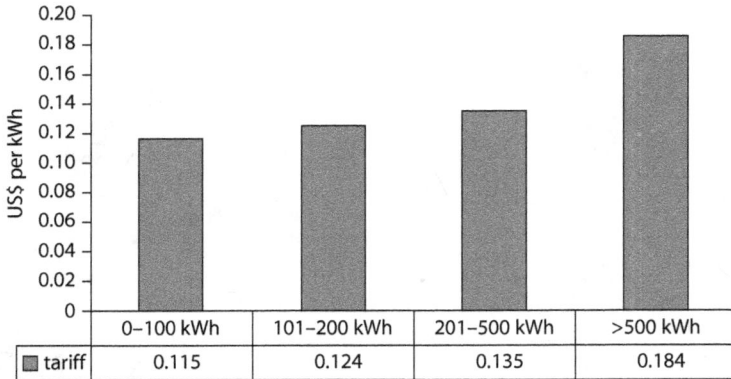

	0–100 kWh	101–200 kWh	201–500 kWh	>500 kWh
tariff	0.115	0.124	0.135	0.184

Source: ONE.
Note: kWh = kilowatt-hour.

Figure 6A.8   Access to Electricity in Morocco, 1995–2010

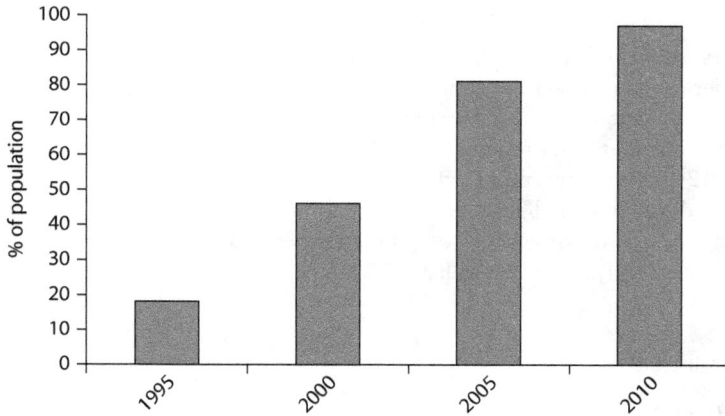

Source: ONE.

Figure 6A.9   Energy Expenditure in Morocco, by Urban and Rural Population, 2001

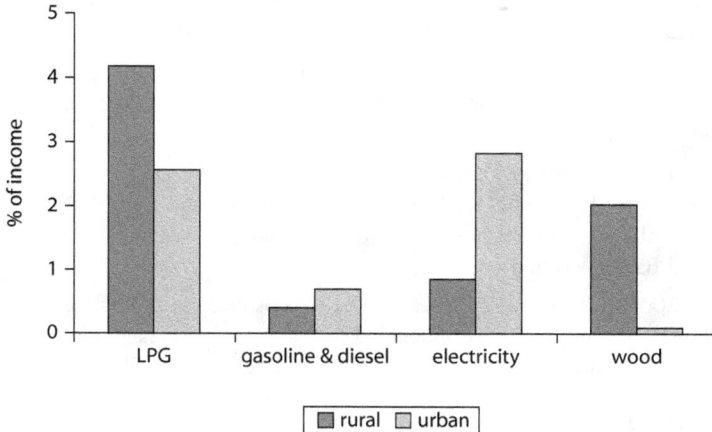

Source: Enquête nationale sur la consommation et les dépenses des ménages, 2001.
Note: LPG = liquefied petroleum gas.

Notes

1. The propane market is deregulated.
2. Key elements of the strategy include (a) diversifying and optimizing the energy mix around reliable and competitive energy technologies; (b) developing the national renewable energy potential with the objective of increasing the contribution of renewables to 10 percent of primary energy demand by 2012; (c) making energy efficiency improvements a national priority; and (d) developing indigenous energy resources by intensifying hydrocarbon exploration activities and developing conventional and nonconventional oil sources (Kelly 2009; World Bank 2009).

References

GIZ (German Agency for International Cooperation). n.d. International Fuel Prices database. GIZ (formerly GTZ), Bonn. http://www.gtz.de/en/themen/29957.htm.

IMF (International Monetary Fund). 2008. "Morocco: 2008 Article IV Consultation." Country Report 304, IMF, Washington, DC.

———. 2010. Retail domestic fuel prices data sheet, IMF, Washington, DC. http://www.imf.org/external/pubs/ft/spn/2010/data/spn1005.csv.

———. 2011. "Morocco: Concluding Statement of the 2011 Article IV Consultation." IMF, Washington, DC.

Kelly, Elaine. 2009. "Pricing for Prosperity: Consumer Price Subsidies in MENA and the Prospects for Reform." Unpublished manuscript, University College London and the Institute for Fiscal Studies, London.

World Bank. 2007. "Program Document for a Proposed Loan to the Kingdom of Morocco for Energy Sector Development Policy Loan." Report 37350-MOR, Sustainable Development Department, Middle East and North Africa Region, World Bank, Washington, DC.

———. 2008a. "Reforming Energy Price Subsidies and Reinforcing Social Protection: Some Design Issues." Energy Sector Management Assistance Program (ESMAP) paper, Report 43173-MNA, World Bank, Washington, DC.

———. 2008b. "Simulating the Impact of Geographic Targeting on Poverty Alleviation in Morocco: What Are the Gains from Disaggregation?" Policy Research Working Paper 4724, World Bank, Washington, DC.

———. 2009. "Social Protection Strategy in Morocco: Targeting and Program Coordination for Better Human Development Outcomes." Concept Note, World Bank, Washington, DC.

CHAPTER 7

Pakistan

Incentives to Energy Subsidy Reforms

The increase in oil and food prices and adverse security developments exacerbated external imbalances in Pakistan. Growing fiscal deficits, owing in large part to increasing energy subsidies and financed by the central bank, fueled inflation. Economic growth was negatively affected by the 2010 floods and the high price of oil, inflation remains persistently high, and budgetary problems are undermining macroeconomic stability.

In October 2008, the authorities embarked on a stabilization program for 2009–10 aimed at restoring economic and financial stability while protecting the poor. This program—supported by the International Monetary Fund (IMF) Stand-By Arrangement approved in November 2008 and extended by nine months in December 2010—envisaged implementing several structural measures as well as strengthening the social safety net (IMF 2009). Implementation of measures to reduce spending on general subsidies in the energy sector has begun.

Despite the fact that Pakistan is well endowed with energy resources,[1] energy imports are still very high. To address the imbalance between energy supply and demand, a series of medium- and long-term measures were discussed in 2010 with the aim of increasing capacity, and a Re 20 billion (US$240 million) energy development fund was launched.

Pakistan aims to bring unscheduled power cuts to an end and to reduce scheduled load shedding (planned power cuts) by one-third. In addition, the government has called on the United States to divert funds from an existing aid package to the energy sector.

The energy crisis is unlikely to be resolved in the short term. In early May 2012, nine independent power producers (IPPs) invoked charges of sovereign default on payment of nearly Re 95 billion. However, the private power-generating companies are themselves caught in the chronic problem of circular debt because each is, in turn, owed money by state-owned power distributors.

Pakistan has resorted to rental power plants (RPPs) as a quick-fix solution to the energy crisis. The aim was to acquire 2,250 megawatts (MW) of RPP capacity by the end of 2009. However, none of the plants was commissioned. Old RPPs (General Electric and Alstom rental plants) could not get gas, resulting in a loss of 286 MW of capacity. The gas shortage forced utilities to use expensive residual fuel oil, leading to greater fuel imports and loss of generation capacity. The financial crisis and the mentioned circular debt problem also played a part as the residual fuel oil-based plants did not receive regular payments, affecting their availability (World Bank 2010a).

Reform Efforts

Fossil Fuels

The Oil and Gas Regulatory Authority was set up under the 2002 power policy to foster competition and increase private investment and ownership in the midstream and downstream petroleum industry. In August 2004, the government also introduced a price differential claim (PDC). The objective was to reimburse oil companies for the subsidy to consumers. The PDC targeted kerosene and diesel. Negative PDC was charged until November 2005 (that is, the fuels were subsidized).

Petroleum prices have been adjusted three times since June 2008, which has led to a significant phaseout of petroleum subsidies. Another series of price increases were implemented in 2010 (see figure 7A.2). As of the end of 2010, diesel and gasoline prices were higher than in the United States even if significantly lower than in the European Union.

Electricity

The power sector regulator, the National Electric Power Regulatory Authority (NEPRA), determines the wholesale tariffs. It is responsible for computing an average scale rate that is different for every distribution company, based on its cost conditions. The government then announces the notified tariff with an implied average scale rate. The difference between the two rates is computed to be the unit subsidy. It is estimated that the government may be subsidizing power to the extent of 22 percent of total cost. Recipients included the Water and Power Development Authority for inter-disco (power distribution companies) differential payments, Karachi Electric Supply Company for tariff differential, and the Power Holding Company for term finance certificate interest payments on circular debt (World Bank 2010a).

At the same time as the petroleum price increases, electricity tariffs were increased by an average of 18 percent effective September 2008. Electricity tariffs were again increased in December 2009 to make up for the shortfall in the October 2009 increase, which was below the 6 percent agreed on with the World Bank and the Asian Development Bank. Subsequently, a 12 percent tariff increase was implemented in January 2010 as well as another 6 percent effective July 2010, as scheduled. Additionally, monthly adjustments are being implemented to recover the fuel costs.

The reforms have not gone without strong objection by domestic business groups and congress members. The government has not been explicit about the reason for the price hikes, often justifying the tariff increases by the recent resurgence in world petroleum prices. Other protests came as a result of price increases for kerosene, which is widely used as cooking fuel by poor Pakistanis, particularly in rural areas (see figure 7A.8).

Poverty Alleviation Measures

Evidence from Household Surveys
Electricity tariffs incorporate a lifeline minimum tariff that shields low-income households from tariff increases (see figure 7A.7). Almost all urban households are connected to the electricity grid, while connection rates for rural areas increase as income rises. Biomass is still used by the majority of households in rural areas, whereas in urban areas its use varies by income (see figure 7A.8). While more than 60 percent of households belonging to the bottom quintile use biomass, only 11 percent of rural houses are connected to the electricity grid. On the other hand, about half of rural households consume kerosene, while few households in large cities consume it. Liquefied petroleum gas is used mainly by richer rural households, as figure 7A.8

shows. Use of natural gas for heating, on the other hand, is mainly an urban consumption item.

Electricity expenditures are more burdensome for poorer households, at least in urban areas. In contrast, the richest quintile spends a larger share on gasoline and diesel (1.7 percent and 3.2 percent in rural and urban areas, respectively) compared with the lowest quintile (0.1 percent and 0.2 percent in rural and urban areas, respectively), as figure 7A.9 shows. The expenditure patterns for kerosene, on the other hand, are neither clearly regressive nor progressive.

Social Safety Nets

The Benazir Income Support Programme (BISP) is the government of Pakistan's main social safety net program to help the poorest families and to cushion the negative effects of price increases. Currently, the government is introducing a new approach to BISP that includes poverty targeting, verification of eligibility criteria, payment case management, and a grievance procedure. The poverty scorecard is based on a proxy means test following international best practice experience (World Bank 2010b).

Prior to the rollout of this new approach, a test phase was launched in selected districts around the country starting in April 2009. The test phase revealed that the poverty scorecard currently in use can be further refined through analysis of more recent household income and consumption data. BISP can collect data or commission the collection of data specifically for the purposes of improving the efficiency of the poverty scorecard.

Key Lessons Learned

Pakistani subsidy reform is an ongoing effort and hence the precise outcome of the reform will not be known for some time. Potential positive results are linked to improving the existing quality of the power services, which has been a source of tension and social unrest.

One of the main drivers behind the recent reform leading to subsidy withdrawals has been the advice and the conditions set by multilateral organizations such as the IMF. The World Bank has also been involved in several energy and social safety net projects.

Annex 7.1 Pakistan Case Study Figures

INCOME LEVEL: Lower-middle income
REGION: South Asia
ENERGY NET IMPORTER/EXPORTER: Net importer
SUBSIDIES: Gasoline, diesel, kerosene, electricity
PHASING OUT SUBSIDIES: Ongoing

Fiscal Burden of Energy Subsidy in Pakistan

Figure 7A.1   Explicit Budgetary Energy Subsidies in Pakistan, 2004–10

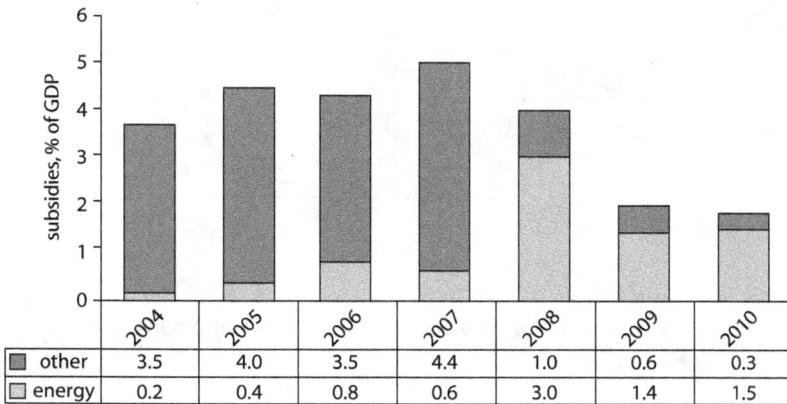

	2004	2005	2006	2007	2008	2009	2010
other	3.5	4.0	3.5	4.4	1.0	0.6	0.3
energy	0.2	0.4	0.8	0.6	3.0	1.4	1.5

Source: Pakistan Ministry of Finance.

Fuel Prices and Road Sector Consumption in Pakistan

Figure 7A.2   Domestic Retail Fuel Prices in Pakistan, 2002–10

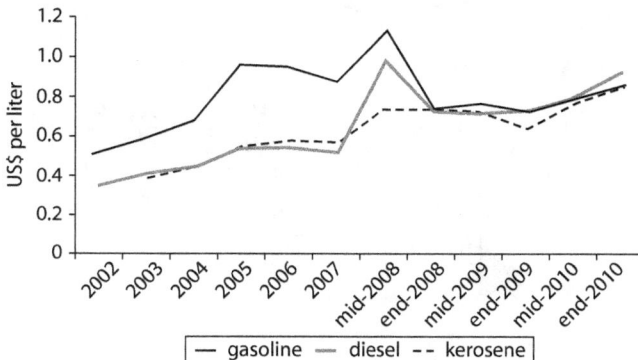

— gasoline — diesel -- kerosene

Sources: Elaboration of data from GIZ n.d.; IMF 2010; and additional data from individual country information.

Figure 7A.3   Road Sector Diesel Consumption in Pakistan, 1998–2008

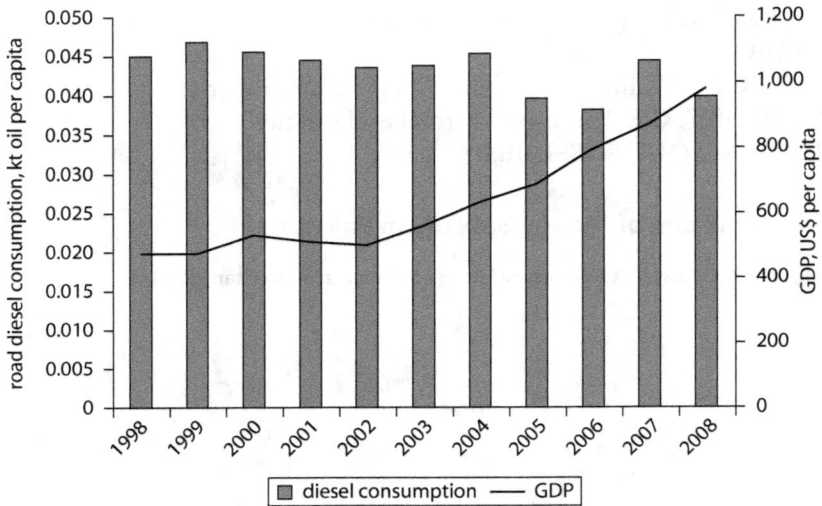

Source: World Bank, World Development Indicators.

Figure 7A.4   Road Sector Gasoline Consumption in Pakistan, 1998–2008

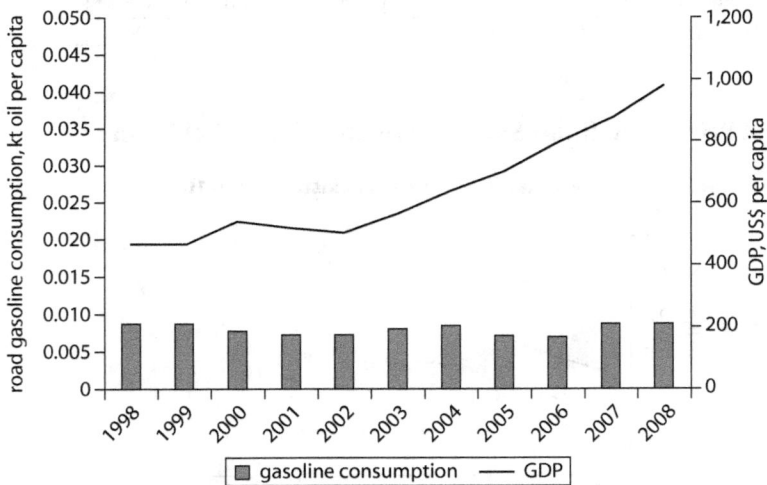

Source: World Bank, World Development Indicators.

Electricity Price and Power Consumption in Pakistan

Figure 7A.5   Electricity Price in Pakistan, 1999–2008

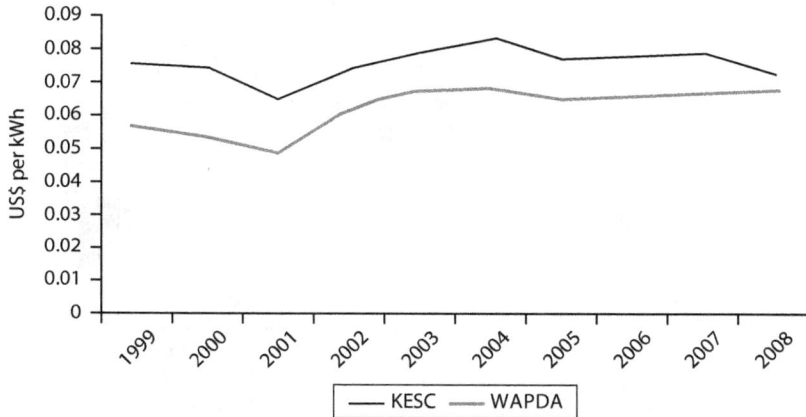

Source: NEPRA various years.
Note: KESC = Karachi Electric Supply Company; WAPDA = Water and Power Development Authority; kWh = kilowatt-hour.

Figure 7A.6   Power Consumption Per Capita in Pakistan, 1998–2008

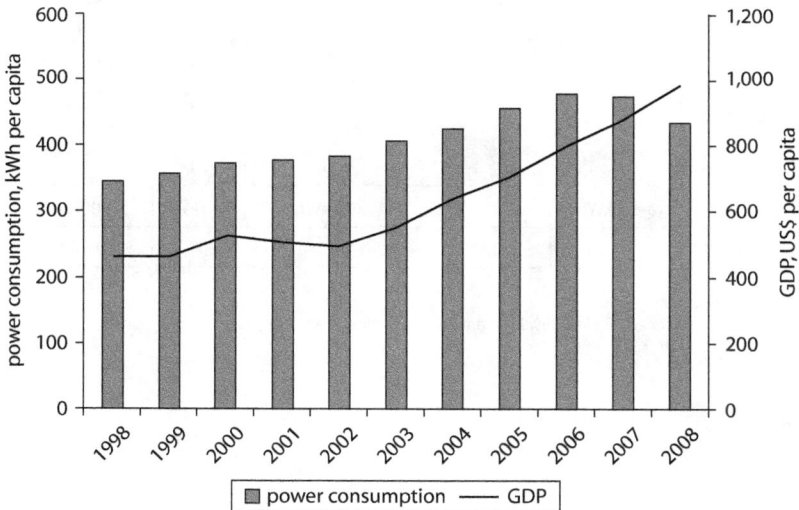

Source: World Bank, World Development Indicators.
Note: kWh = kilowatt-hour.

Poverty Impact Evidence from Household Surveys in Pakistan

Figure 7A.7   Electricity Block Tariffs in Pakistan, 2010

a. WAPDA

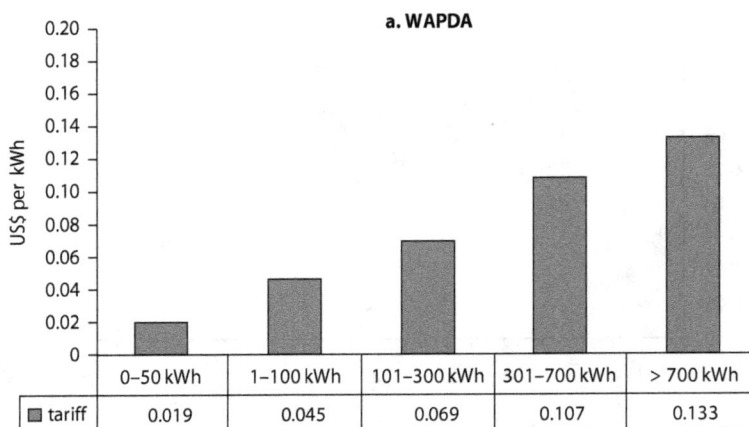

	0–50 kWh	1–100 kWh	101–300 kWh	301–700 kWh	> 700 kWh
■ tariff	0.019	0.045	0.069	0.107	0.133

b. KESC

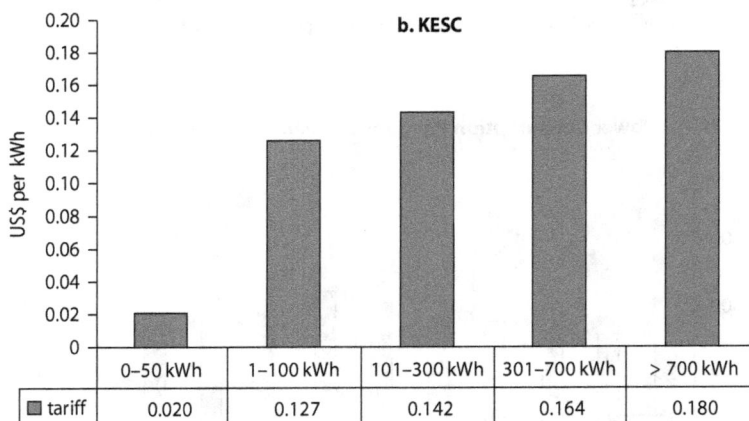

	0–50 kWh	1–100 kWh	101–300 kWh	301–700 kWh	> 700 kWh
■ tariff	0.020	0.127	0.142	0.164	0.180

Source: NEPRA various years.
Note: WAPDA = Water and Power Development Authority; KESC = Karachi Electric Supply Company;
kWh = kilowatt-hour.

Figure 7A.8   Household Use of Energy Sources in Pakistan, by Income Quintile, 2005

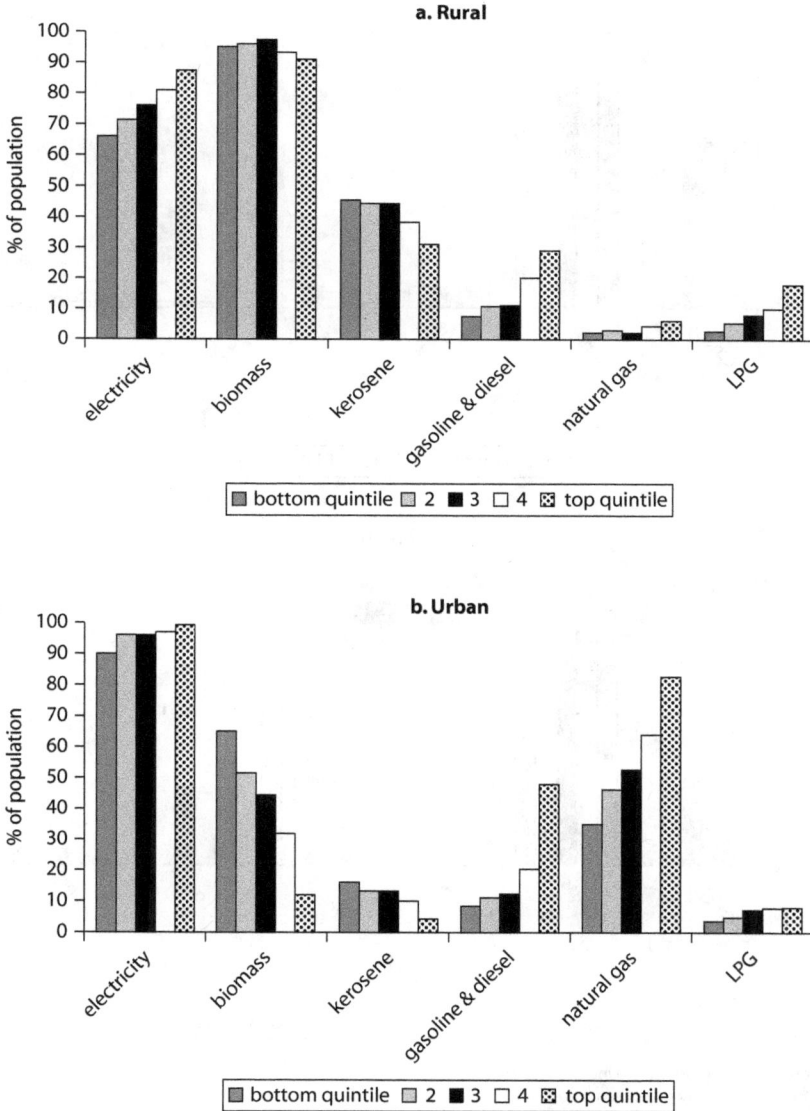

a. Rural

b. Urban

Source: Pakistan Household Income and Expenditure Survey, 2005.
Note: LPG = liquefied petroleum gas.

Figure 7A.9   Household Energy Expenditure in Pakistan, by Income Quintile, 2005

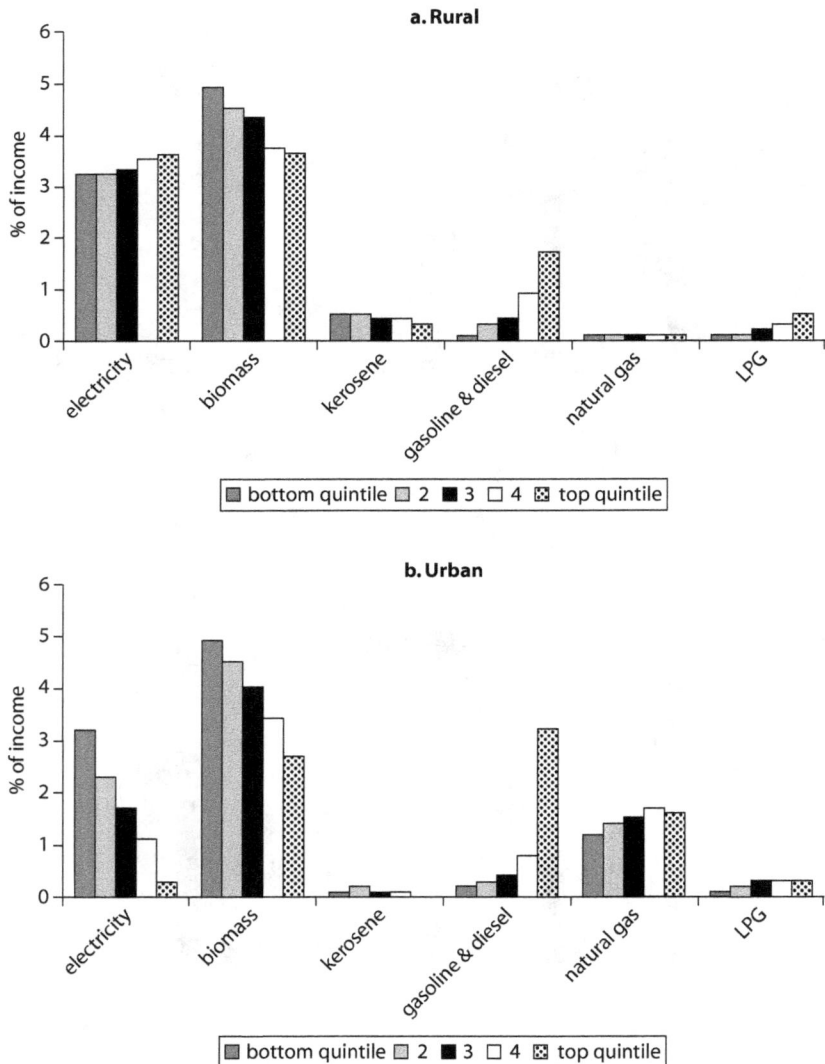

a. Rural

bottom quintile ☐ 2 ■ 3 ☐ 4 ▧ top quintile

b. Urban

bottom quintile ☐ 2 ■ 3 ☐ 4 ▧ top quintile

Source: Pakistan Household Income and Expenditure Survey, 2005.
Note: LPG = liquefied petroleum gas.

Note

1. About 937 million barrels of oil have been discovered, of which 354 million barrels remain unexploited, and the coal reserves are estimated at 185 billion. Pakistan also has a hydropower potential of about 45,000 megawatts (MW), of which only about 6,450 MW has been developed.

References

GIZ (German Agency for International Cooperation). n.d. International Fuel Prices database. GIZ (formerly GTZ), Bonn. http://www.gtz.de/en/themen/29957.htm.

IMF (International Monetary Fund). 2009. "Pakistan: Article IV Consultation." Country Staff Report 09/123, IMF, Washington, DC.

———. 2010. Retail domestic fuel prices data sheet, IMF, Washington, DC. http://www.imf.org/external/pubs/ft/spn/2010/data/spn1005.csv.

NEPRA (National Electric Power Regulatory Authority). Various years. Annual Reports. NEPRA, Islamabad, Pakistan.

World Bank. 2010a. "Impact of the Global Financial Crisis on Investments in South Asia's Electric Power Infrastructure." Report 56849-SAS, Energy Sector Management Assistance Program and World Bank, Washington, DC.

———. 2010b. "Pakistan: Process Evaluation of Benazir Income Support Programme (BISP) Scorecard-Based Poverty Targeting under the Test Phase." Report on Findings, GHK Consulting Limited for the World Bank and the Government of Pakistan, Faisalabad, Pakistan.

PART 2

Group B Countries:
Net Energy Importer and High Income

Macroeconomic and Social Challenges

- All countries are characterized by an increasing level of income, as displayed by a buoyant growth in gross domestic product (GDP) per capita (see figure P2.1) as well as decreasing income inequality over time.
- Chile stands out as the richest country but also as the country with the highest level of income inequality within Group B. The Latin American countries are characterized by the highest levels of inequality. Among them, Peru, the poorest country in Group B, was most successful in reducing inequalities (see figure P2.2).
- The majority of Group B countries are characterized by a decreasing or stable budget and public debt over time (see figures P2.3 and P2.4), with the notable exception of the Dominican Republic. Its budget moved from a small fiscal surplus in 1998 to a fiscal deficit of 3 percent of GDP in 2008 and a deterioration of the public debt. In contrast, Chile recorded the highest reduction in public debt and an increase in the budget primary surplus.

Fossil Fuel Dependence

- All countries with the exception of the Dominican Republic increased the percentage of electricity generated from fossil fuels as well as energy net imports (see figures P2.5 and P2.6).
- The Dominican Republic relies almost entirely on fossil fuels and is characterized by the highest net imports of energy.

Income and Inequality Trends for Group B

Figure P2.1   GDP Per Capita, Group B Countries, 1998–2008

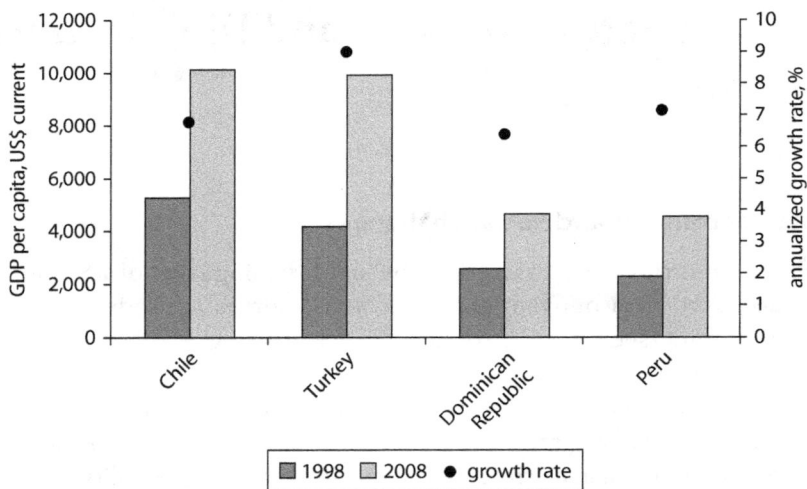

Source: World Bank, World Development Indicators.

Figure P2.2   Gini Index, Group B Countries, 1998–2008

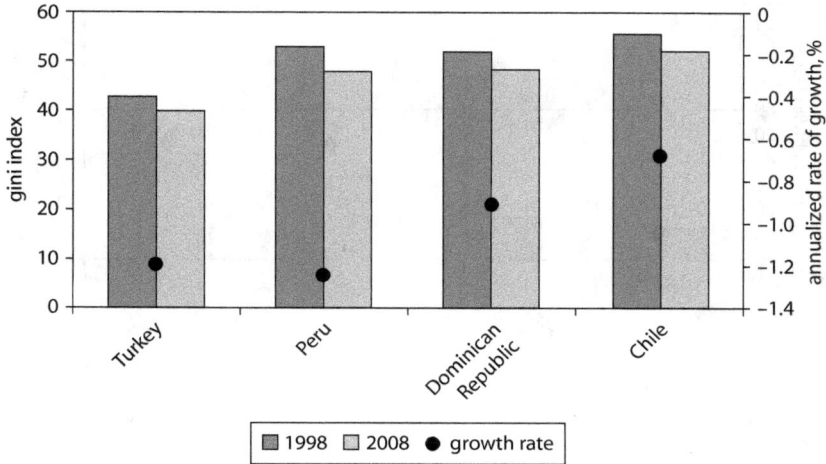

Source: World Bank, World Development Indicators.
Note: The Gini index measures the extent to which the distribution of income (or consumption expenditure) among individuals or households within an economy deviates from a perfectly equal distribution. A Gini index of 0 represents perfect equality, while an index of 100 implies perfect inequality (World Bank, *World Development Indicators*).

Fiscal Indicators for Group B

Figure P2.3   General Government Net Lending or Borrowing, Group B Countries, 1998–2008

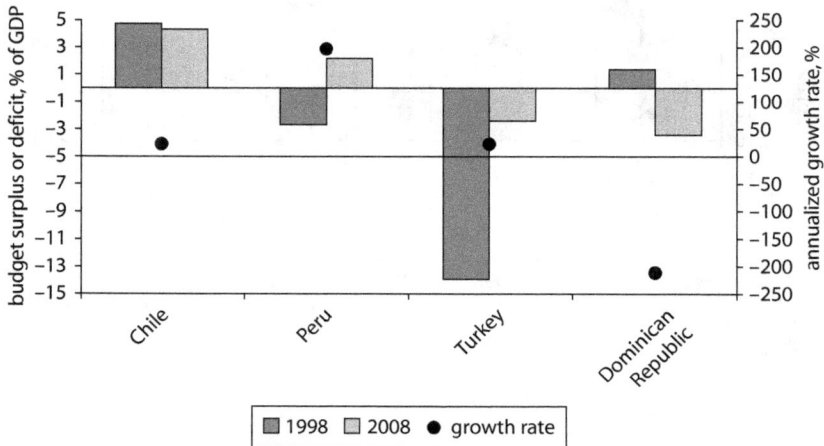

Source: IMF reports, various years.

Figure P2.4   General Government Gross Debt, Group B Countries, 1998–2008

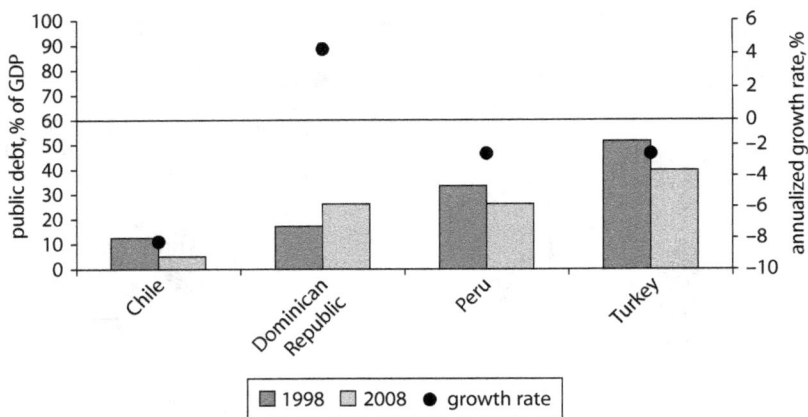

Source: IMF reports, various years.

Fossil Fuel Dependence for Group B

Figure P2.5   Electricity Production from Fossil Fuels, Group B Countries, 1998–2008

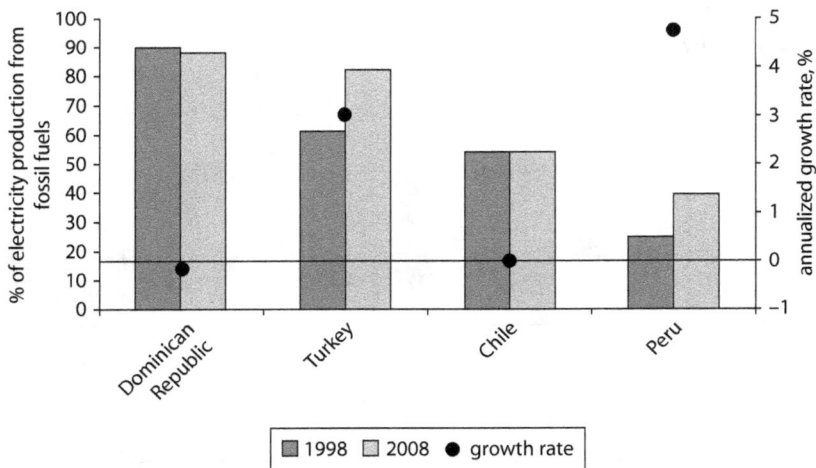

Source: World Bank, World Development Indicators.

Figure P2.6   Energy Net Imports, Group B Countries, 1998–2008

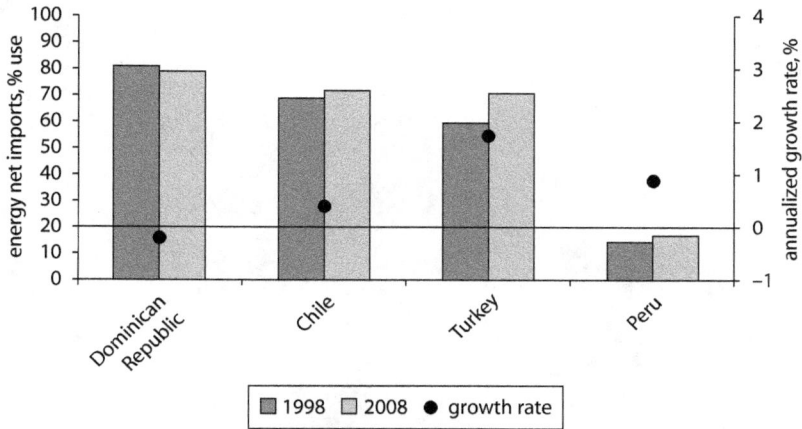

Source: World Bank, World Development Indicators.

CHAPTER 8

Chile

Incentives to Energy Subsidy Reforms

Chile has significant hydroelectric resources, contributing to 45 percent of its electricity supply. Nevertheless, with little indigenous production of fossil fuels, Chile imports oil, natural gas, and coal and, until the arrival of liquefied natural gas (LNG) in June 2009, it also imported gas from Argentina. In the past decade, Chilean energy policy has been oriented toward reducing the country's vulnerability to supply shocks and dependence on energy imports, which have grown on the back of increased energy consumption. Reducing reliance on hydropower has also been sought, as volatile weather patterns exposed Chile to electricity shocks in the 1990s. Since 2004, when Argentina cut its natural gas supplies to Chile sharply in response to a domestic energy crisis, the Chilean government has sought to diversify fuels and suppliers. The ongoing effort has included incentives to build hydroelectric and coal-based projects as well as programs to encourage the use of renewable energy sources.

In 2008, Chile again lost most of its gas imports from Argentina at a time when Chile was severely affected by drought. It substituted costly diesel oil to run power stations originally built to run on natural gas. The LNG port terminal at Quintero (in the central region) started receiving gas shipments from Trinidad and Tobago in June 2009. A second plant located in Mejillones, in the north of the country (a joint venture between GDF Suez and Codelco, the state copper mining company), began operating in July 2010 and will mostly be used to serve the mining industry in the northern regions.

Reform Efforts

Fossil Fuels

The national oil company (*Empresa Nacional del Petróleo;* ENAP) dominates not only oil extraction but also refining as well as pipeline transport in partnership with other companies. Since the liberalization of its electricity market, Chile has had no subsidies or price capping for fuels. Prices for petroleum-based fuels are set by the refiner and are based on cost reflection throughout the distribution chain.

However, two fuel stabilization funds have been established to reduce the price volatility of imported fossil fuels:

- The Fuel Price Stabilization Fund (*Fondo de Estabilización de Precios de los Combustibles;* FEPC) from 1991 to 2005, for oil
- The Petroleum Price Stabilization Fund (*Fondo de Estabilización de Precios del Petróleo;* FEPP), from 2005 to 2010, for liquefied petroleum gas, LNG, gasoline, diesel, and kerosene.

The funds maintained the price of fuel imports within a price band to match recent average import price levels. The import parity price was calculated each week and compared to the band limits. This determined whether a credit or a tax would be applied to end-user prices to keep the price within the 12.5 percent band in the case of FEPP (IEA 2009). Although similar to the FEPP, the FEPC was characterized by a smaller margin of fluctuation (5 percent). Also, calculation of the import parity price was not based on the cost, insurance, and freight price of crude oil but instead on the standard West Texas Intermediate price. Neither of the funds was self-financing, and both required capital injections by the government (see table 8.1 below).

FEPC ceased operating in 2010 and was replaced by SIPCO (*Sistema de Protección ante Variaciones de Precios de Combustibles*). With the underlying objective of reducing the fiscal burden of the previous schemes, the new regime introduced two major components to protect against price volatility. First, the government reduces or increases the fuel tax rate in response to reductions or increases of the international price of oil, reinstating the 12.5 percent band that was utilized during FEPP (1991–2005). Second, if the spot price is higher than the reference

Table 8.1   Fiscal Cost of Fuel Stabilization Funds in Chile, 2000–09
US$, millions

	2000–05 (FEPP)	2006–09 (FEPC)
Direct fiscal cost	353	824
Lower revenues from VAT	81	236
Lower revenues due to reduction of fuel tax	n.a.	850
Total	434	1,910

Source: Ministry of Finance of Chile.
Note: FEPP = Petroleum Price Stabilization Fund; FEPC = Fuel Price Stabilization Fund; VAT = value added tax.
n.a. = not applicable.

price, an insurance compensation mechanism is triggered. Both the compensation and the insurance premiums are transferred to the variable component of the fuel tax (Larraín 2010).

Electricity

The power sector was reformed during the 1980s as part of the government's countrywide reorganization of the economy. The process of reforming the Chilean power sector was implemented through legal and institutional changes. Private participation was encouraged by establishing new investor-financed enterprises to purchase existing facilities or to construct new facilities. Between 1983 and 1989, the government privatized most of the generation, transmission, and distribution segments through local and international investors and created a mandatory power pool administered by the system operator, CDEC-SIC (*Centro de Despacho Económico de Carga del Sistema Interconectado Central*) (Vagliasindi and Besant-Jones, forthcoming).

Tariffs for residential customers varied in the range of US$0.08 to US$0.10 per kilowatt-hour (kWh) until 2004—which was generally affordable—but then increased quickly to US$0.15 per kWh in 2007 and to almost US$0.20 in 2008 mainly because of increased generation costs (see figure 8A.4).

The government has sought to strengthen the sector's policy and regulatory governance by creating a Ministry of Energy, operational since March 2010. The existing regulator, the National Energy Commission (*Comisión Nacional de Energía*; CNE), retains its responsibilities, whereas the new energy ministry aims at reaching long-term policy goals such as increasing capacity, reducing exposure to supply shocks, and implementing energy efficiency programs. Energy efficiency is already an integral part of Chilean energy policy, evidenced by the substantial budget increases given to the National Energy Efficiency Program (*Programa País de Eficiencia Energética*) since it began operating in 2006. According to official sources, energy efficiency programs achieved a 2.6 percent annualized reduction in energy demand in the main electricity grid. In April 2010, the Ministry of Energy presented a strategy to increase energy efficiency by 2020, known as ChileE3.

The electricity bills of low-income residential users are subsidized in certain cases. The subsidy is applied only if the tariff increases by 5 percent or more within a period of six months or less (IEA 2009). A transitory subsidy in the form of a discount on the electricity bill reduces the immediate impact of a tariff increase on poor households.

Chile has further enhanced rural electrification with government support because it is typically unprofitable for private power companies to invest in rural areas. Chile's rural electrification fund was launched in 1992, resulting in the creation of a special mechanism (the fund) linking subsidies to output targets. The central government allocates the subsidy fund to the regions based on the number of unelectrified households and the progress each region has made in the development of renewable energy projects in the preceding year. The subsidies are competitively allocated as one-time direct subsidies to private distribution companies to cover their investment costs (see table 8.2). Local operators apply for the subsidies by submitting details of their proposed projects, which then are scored against a checklist of objective criteria, including cost-benefit analysis, operator investment commitment, and social impact. By 2003, Chile had reached 97 percent electrification coverage.

Off-grid connections, mostly operated by local distributors, are typically more expensive to operate. The tariffs are set through formal agreements between the mayor and the legal representative of the local electricity provider (IEA 2009). Since 2009, the government has started subsidizing off-grid electrification so that same tariff rates apply to all residential electricity users.

Poverty Alleviation Measures

Evidence from Household Surveys

Energy expenditures for all sources of fuels are more burdensome for poorer households and account for a larger share of their household expenditures (see figure 8A.6). Spending on electricity is the most regressive: the bottom quintile spends a budget share of almost 5 percent on electricity while the richest quintile spends only 1.5 percent. The same pattern applies for different types of fuels, with the bottom quintile

Table 8.2   Rural Electrification Program in Chile, 1992–2000

Year	Electrified households (number)	Total subsidy volume (Ch$, millions)	(US$, millions)	Average subsidy[a] (US$, thousands)	Subsidy as percentage of total investment
1992	8,442	2,668	7.5	890	78
1993	9,123	3,378	8.4	918	71
1994	8,370	2,655	6.3	754	67
1995	17,933	7,749	20.7	1,157	70
1996	19,053	9,722	23.7	1,245	65
1997	19,107	10,813	25.9	1,356	64
1998	20,427	13,191	28.9	1,416	61
1999	13,625	7,927.6	15.8	1,159	62
2000	13,901	8,113.9	15.3	1,102	66
Total	**129,981**	**66,218**	**153**	**9,997**	—

Source: Fischer and Serra 2003.
Note: — = not available.
a. One-time subsidy per private distribution company.

spending a larger budget share on petroleum fuels (3.2 percent) than the highest quintile (1.8 percent). The expenditure patterns for natural gas and LPG are also clearly regressive.

The government of Chile has developed a robust social database system based on extensive collection of household data, which enables it to successfully identify households by income for social program targeting. The Index of Socioeconomic Characterization (*Ficha de Caracterización Socioeconómica*; Ficha CAS) is a two-page form used for collecting detailed household information such as educational levels, incomes, occupations, and so on. The form allows the government to determine household eligibility for a wide range of government programs. The Ficha CAS is updated every three years and assigns points to households on the basis of the information collected. The points give an indication of how much assistance, if any, the family would receive.

One of the advantages of using such proxy means testing for many different programs is that the cost of testing is reduced. The cost of one interview is about US$8.65 per household. Given that the fixed administrative costs are spread across several programs, the Ficha CAS is cost-effective. In 1996, administrative costs represented a mere 1.2 percent of the benefits distributed using the Ficha system. To illustrate, if the administrative costs of the Ficha system were to be borne by water subsidy programs alone, they would represent 17.8 percent of the value of the subsidies (World Bank 2008).

Key Lessons Learned

Chile's energy sector reform success and sustainability are most impressive considering that it imports 80 percent of its primary energy and has few indigenous fossil fuel resources. At the same time, this leaves Chile vulnerable to price volatility and supply interruptions.

Chile has recently faced two crises to its energy supply for power generation to which the reformed sector responded well, partly because the system's generation capacity reserve margin remained substantial (at around 60 percent) from 1999 onward. The first crisis was during 1998 and 1999—soon after the reforms were implemented—when Chile suffered one of the worst droughts on record and hydropower generation dropped by about a third. Despite an increase in thermal power generation, Chile experienced some power rationing at that time. The second crisis was the reduction of gas supplies from Argentina from May 2005 onward, stemming from rising Argentinean demand for residential winter heating and other uses. Chile had relied on these imports to cover around 30 percent of its total power generation. Consequently, the amount of power generated from natural gas in Chile declined sharply, and Chile had to switch gas-fired generation plants to liquid petroleum fuels at a substantial increase in operational expenses. In response to this risk to imported gas by pipeline, Chile inaugurated in October 2009 its first natural-gas regasification plant for importation of LNG as part of a strategy to diversify energy sources.

Chile is widely regarded as a successful case of energy sector reform because it was one of the first countries in the region to successfully liberalize its electricity market. The subsidiary role of the state in the electricity sector is one of the key lessons for countries undergoing reform, as is the separation of functions in energy sector management. The primary focus of the state has been to direct long-term energy policy. More recently there has been an increased role for the government regarding the security of supply and in establishing the Ministry of Energy, but competition and private sector participation in the electricity sector rule—and pricing via the principle of supply and demand were always kept as guiding principles.

One of the areas that Chile has been looking to address more carefully is diversification of its energy mix and its energy suppliers in order to decrease import dependence and avoid a single-supplier scenario. There is an especially strong potential for hydropower development in the south, which Chile hopes to further exploit, but imported oil, coal, and natural gas will most likely remain the primary energy sources for the years to come.

Annex 8.1 Chile Case Study Figures

INCOME LEVEL: High-income OECD (Organisation for Economic
Co-operation and Development)
REGION: Latin America and the Caribbean
ENERGY NET IMPORTER/EXPORTER: Net importer
SUBSIDIES: Electricity
PHASING OUT SUBSIDIES: Successful

Fuel Prices and Road Sector Consumption in Chile

Figure 8A.1   Domestic Retail Fuel Prices in Chile, 2002–10

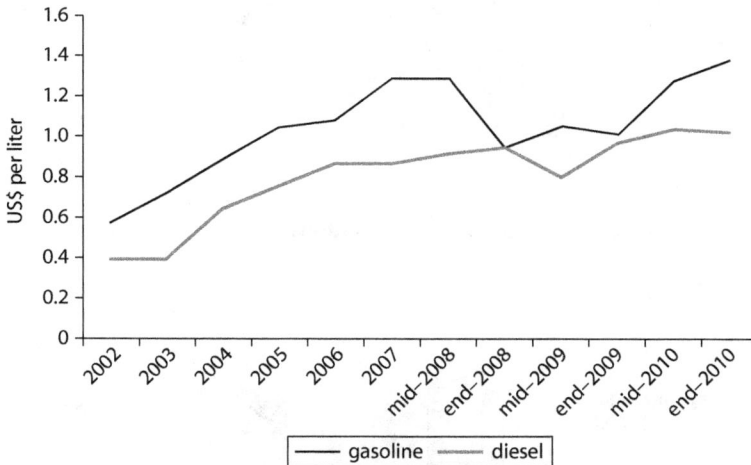

Source: CNE.

Figure 8A.2   Road Sector Diesel Consumption in Chile, 1998–2008

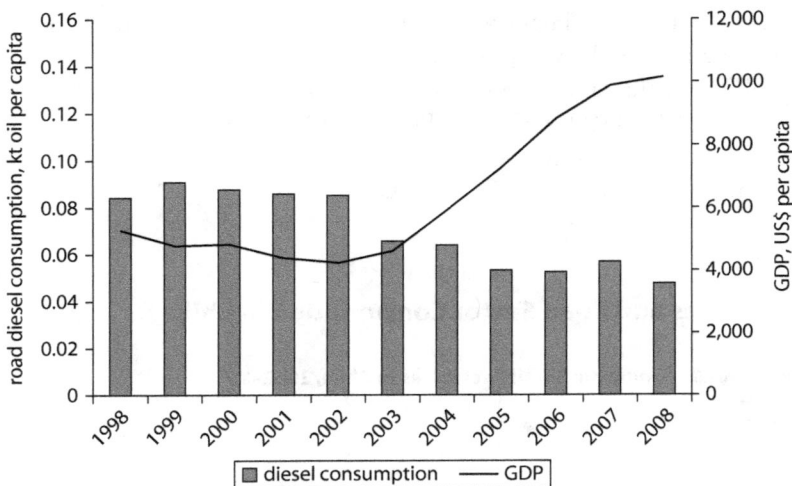

Source: World Bank, World Development Indicators.

Figure 8A.3   Road Sector Gasoline Consumption in Chile, 1998–2008

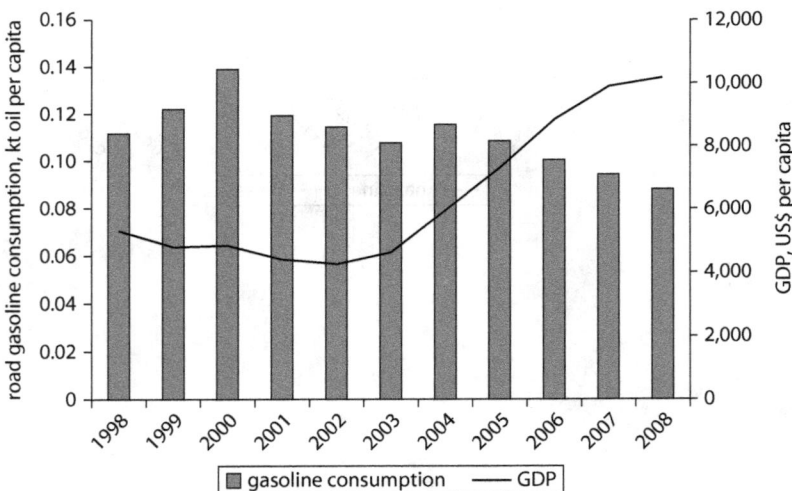

Source: World Bank, World Development Indicators.

Electricity Price and Power Consumption in Chile

Figure 8A.4   Electricity Price in Chile, 1998–2008

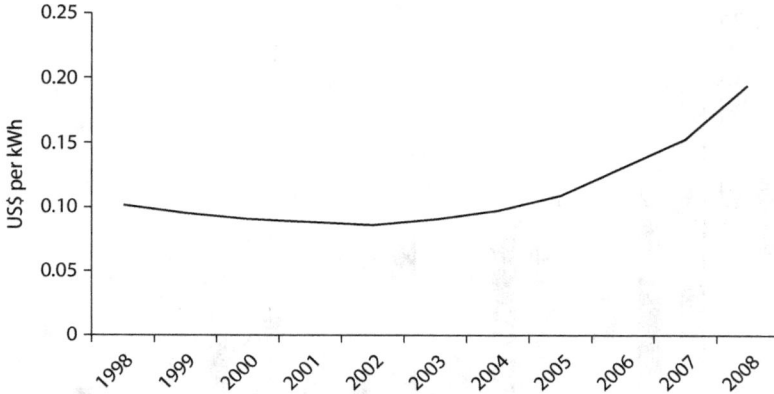

Source: CNE.
Note: kWh = kilowatt-hour.

Figure 8A.5   Power Consumption Per Capita in Chile, 1998–2008

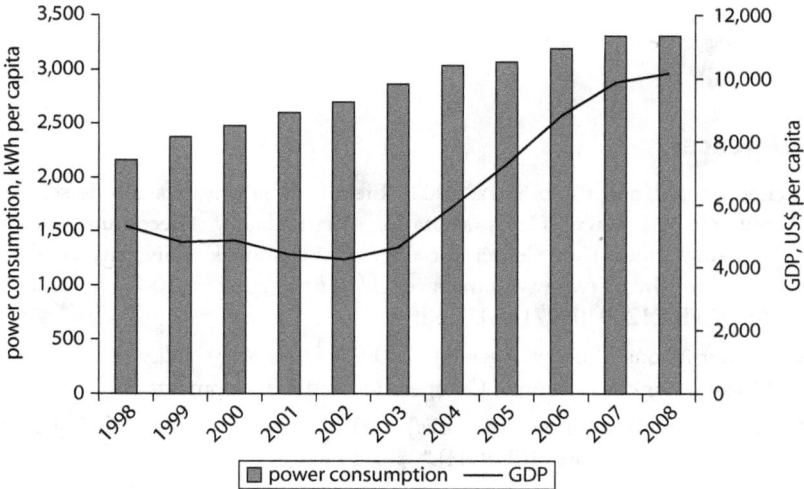

Source: World Bank, World Development Indicators.
Note: kWh = kilowatt-hour.

Poverty Impact Evidence from Household Surveys in Chile

Figure 8A.6   Energy Expenditure in Chile, by Income Quintile, 2007

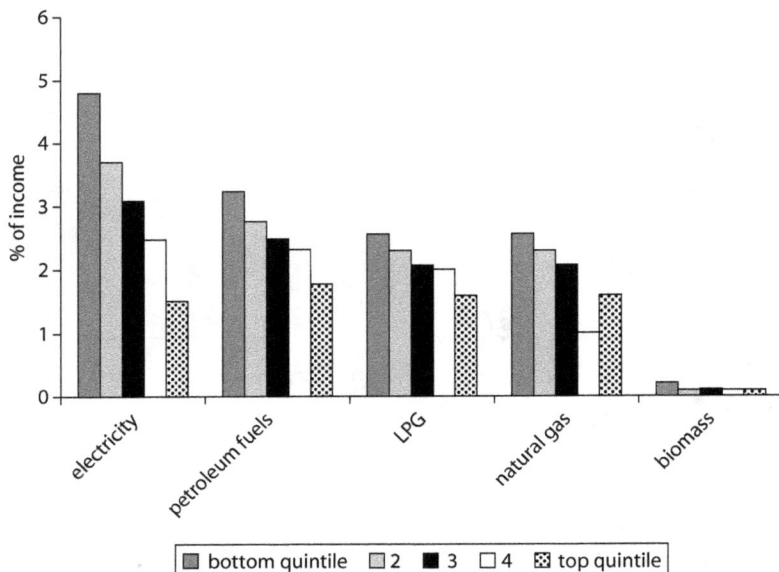

Source: ONE 2007.
Note: LPG = liquefied petroleum gas.

References

Fischer, Ronald, and Pablo Serra. 2003. "Efectos de la privatización de servicios públicos en Chile: Casos sanitario, electricidad y telecomunicaciones." Working Paper 186, Center for Applied Economics, University of Chile, Santiago. http://www.webmanager.cl/prontus_cea/cea_2004/site/asocfile/ASOCFILE120040527142057.pdf.

IEA (International Energy Agency). 2009. *Chile Energy Policy Review*. Paris: Organisation for Economic Co-operation and Development/IEA.

INE (Instituto Nacional de Estadística). 2007. "Encuesta Nacional de Ingresos y Gastos de los Hogares (ENIGH)."

Larraín, Felipe. 2010. "Sistema de Protección ante Variaciones de Precios de Combustibles (SIPCO)." Presentation of the Minister of Finance of Chile, Santiago, Chile.

Vagliasindi, Maria, and John Besant-Jones. Forthcoming. *Power Market Structure: Revisiting Policy Options*. Directions in Development Series. Washington, DC: World Bank.

World Bank. 2008. "Reforming Energy Price Subsidies and Reinforcing Social Protection: Some Design Issues." Energy Sector Management Assistance Program (ESMAP) paper, Report 43173-MNA, World Bank, Washington, DC.

CHAPTER 9

The Dominican Republic

Incentives to Energy Subsidy Reforms

Until recently, the large fiscal burden imposed by the energy sector through direct and indirect subsidies crowded out expenditures, while blackouts and demand for alternative self-generated electricity raised production costs. The Dominican Republic's power sector stands out from the other countries in the Latin America and Caribbean region because of its high nontechnical losses, amounting to about 40 percent. (More than 30 percent of electricity is used illegally and for free.) Technical losses, for their part, are in line with regional averages. The Dominican Republic is facing multiple challenges, including restrictive contracts between generators and distributors with non-cost-reflective energy tariffs; a weak electricity tariff structure (due to the absence of an automatic adjustment mechanism); a badly designed subsidy structure; low operating efficiency of the distribution companies; and weak financial planning.

In order to extend the hours of energy provision, the government established an extensive electricity subsidy program in 2001 called the Blackout Reduction Program (*Programa de Reducción de Apagones*; PRA), thereby trying to avoid public protests, but it was not until after the banking and macroeconomic crisis in 2004 that the government decided to freeze consumer rates. In 2008, the electricity transfer payments accounted for RD$41.9 billion (or US$1.2 billion), equivalent to 2.7 percent of gross domestic product (GDP) (see figure 9A.1) (IMF 2011).

Reform Efforts

Sector reform started in the early 1990s with the opening up of generation to private investors. A central step was the unbundling of the Dominican Electricity Corporation (*Corporación Dominicana de Electricidad*; CDE) into two generation companies (Itabo and Haina) and three regional electricity distribution companies (EDEs) (EdeNorte, EdeSur, and EdeEste) in 1997. These companies were subsequently privatized through the sale of 50 percent shares (including managerial control), with the remaining shares held by the Endowment Fund of the Reformed Enterprises (*Fondo Patrimonial de las Empresas Reformadas*; FONPER).

For decades, the electricity sector has provided substandard service, and consumers experienced frequent power cuts. Reducing the high level of commercial losses in the distribution system and improving the poor quality of power supply led to a vicious circle, as poor quality of service, customer dissatisfaction, and high tariffs induced theft through illegal connections and nonpayment of electricity bills by businesses and households. This, in turn, has left the distribution companies without the resources to make the necessary improvements. The problem has been compounded by the difficulty of cutting off supplies for overdue bills, together with the impunity with which consumers reconnect to the grid illegally.

Electricity tariffs in the Dominican Republic average about US$0.18 per kilowatt-hour (kWh) and are among the highest in Latin America, mostly because of the country's reliance on imported oil for power generation. To some extent, the high tariff is due to the increase in international oil prices, but it is also an effect of the unfavorable terms of the country's negotiated power purchase agreements, including a certain risk premium reflecting the repeated power sector crises and low cash recovery, as recounted below:

- Fossil fuel prices skyrocketed from US$0.40 per liter in 2002 to the peak of about US$1.40 in mid-2008 (see figure 9A.2).
- Road sector fuel consumption, partly due to the period of recession (particularly in the case of diesel) was significantly reduced and did not

resume growth despite the recovery of GDP growth in 2005 (see figures 9A.3 and 9A.4).

- When fuel prices rose in March 2003 because of the Iraq war, the government did not adjust the electricity tariff for residential customers' first blocks (up to 700 kWh) and created a stabilization fund to cover the difference.
- In late 2007, the government decided to freeze retail tariffs despite the increase in oil prices (fuel oil rose from US$40 per barrel in May 2007 to US$56 per barrel in September 2007). Appropriate adjustments would have required an average increase of about 15 percent. The authorities feared that a further increase would result in a larger non-payment of electricity bills, thereby counteracting the ongoing efforts to increase the cash recovery index.
- Authorities increased electricity tariffs by 6.4 percent in June 2009 and by 11 percent in December 2010 (see figure 9A.5).

Poverty Alleviation Measures

Evidence from Household Surveys
The existence of an increasing block tariff for electricity (see figure 9A.7) does not offer real protection to poor people in light of the low rate of metering, which is equivalent to one-third of consumption. Half of electricity consumption is not metered, and 20 percent of electricity consumers only pay the connection fee (see figure 9A.8).

Electricity accounts for the lion's share of consumption of the average household in the Dominican Republic—representing 3.5 percent of household income (70 percent of the total household energy expenditure)—followed by natural gas, mainly used for cooking and heating, which adds a 1.4 percent burden to the budget (see figure 9A.9).

The World Bank has played a major role to support social protection mechanisms as well as efficiency of public spending, among other goals (World Bank 2009). The Inter-American Development Bank has also been helping the authorities to implement better targeting of liquefied petroleum gas (LPG) subsidies (IDB 2009).

The PRA, as a geographic subsidy mechanism, does not display sufficient targeting effectiveness to assist the households who need the relief the most and is thus not the proper scheme to streamline electricity subsidies. In the zones covered by the PRA, there are small companies, stores, and homes with medium to high incomes receiving the same reduced rates as homes with much lower household incomes. In fact, only about half of the households served by the PRA can be categorized as low income (IDB 2009; World Bank 2009). The categories of consumers that benefit the most are companies and households that consume the most energy, which are typically not the poorest ones. The energy use habits of end users in these areas are markedly different from those whose electricity costs more because there is inefficient use of electricity through indiscriminate use of air conditioners and other energy-intensive equipment. Furthermore, this system of subsidies generated a perverse incentive for businesses to move to the areas covered by the PRA in order to be covered by energy subsidies and reduce their overhead costs.

Social Safety Nets

According to data about the 2.28 million households in the Unified System of Beneficiary Identification (*Sistema Único de Beneficiarios*; SIUBEN)—the country's targeting instrument program—only about 1.2 million have valid contracts with one of the EDEs, while the rest of the households (1.1 million, or 48 percent) don't have regulated electricity access. The users residing in the zones covered by the PRA do not have any incentive to switch into a formal contract with the distributing companies.

The Dominican Republic began reducing LPG subsidies in September 2008. LPG is used by most consumers as a home fuel and even for driving in some vehicles. Only about 800,000 families identified as low-income by SIUBEN will be receiving a reduced price for a maximum of six gallons of LPG per month.

The new subsidization scheme replacing universal subsidies is called BonoLuz. The broader goal of BonoLuz is to improve the government's fiscal balance and administration by reducing subsidies and helping to recover electricity costs, but it is largely based on redesigning the conditional cash transfer program (Solidaridad) and improving the targeting instrument for social spending through SIUBEN, which will be designated as the primary mechanism for determining the distribution of subsidies (see figure 9A.10). BonoLuz uses coupons for the poorest consumers to claim a subsidy for the use of the first 100 kWh. In this case, the utility companies would deduct 100 kWh from the electricity bill of particular households. Accordingly, the electricity block tariff may be discontinued. In October 2010, the number of clients amounted to half of the increase of clients targeted by the end of December 2011 (IMF 2011).

Key Lessons Learned

Subsidy reforms in the Dominican Republic are still ongoing, and hence clear outcomes are difficult to assess. However, some lessons emerge:

- One of the key emerging lessons from the Dominican Republic case is that geographic subsidies in urban and periurban areas may not always work because they tend to create perverse incentives and attract commercial use. This situation, as a whole, also promotes energy overconsumption and inefficiency.

- Removing geographic subsidies and directing the savings toward covering more households that are in actual need of assistance is a step in the right direction. The challenge that the new BonoLuz program will have is that it will only be able to cover those families possessing Solidaridad cards—only about a third of those who were receiving the benefits of the PRA (and thus have not been paying for electricity). The rest have yet to become registered to start receiving the benefits of BonoLuz.

- The main challenge for the electricity distribution companies is their dependence on the government to cover their operating costs. As of 2006, increases in tariffs and the cost recovery index have been accompanied by better quality of the service delivered, which has made the improvements in quality socially acceptable.

Annex 9.1 Dominican Republic Case Study Figures

INCOME LEVEL: Upper-middle-income
REGION: Latin America and the Caribbean
ENERGY NET IMPORTER/EXPORTER: Net importer
SUBSIDIES: Electricity, LPG
PHASING OUT SUBSIDIES: Successful, ongoing

Fiscal Burden of Energy Subsidy in the Dominican Republic

Figure 9A.1   Explicit Budgetary Energy Subsidies in the Dominican Republic, 2004–10

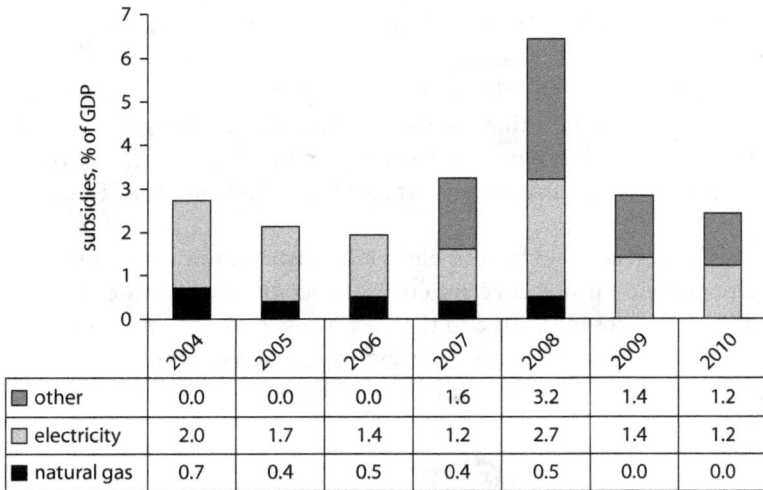

	2004	2005	2006	2007	2008	2009	2010
■ other	0.0	0.0	0.0	1.6	3.2	1.4	1.2
▢ electricity	2.0	1.7	1.4	1.2	2.7	1.4	1.2
■ natural gas	0.7	0.4	0.5	0.4	0.5	0.0	0.0

Sources: IMF 2011; IDB 2009.

Fuel Prices and Road Sector Consumption in the Dominican Republic

Figure 9A.2   Domestic Retail Fuel Prices in the Dominican Republic, 2002–10

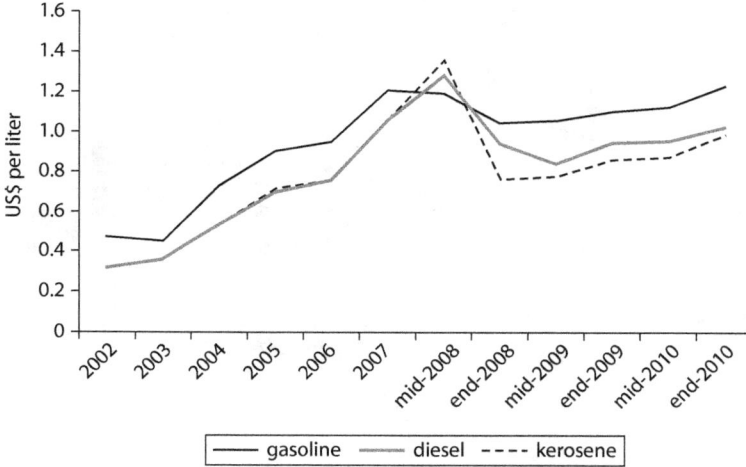

Source: SEIC 2011.

Figure 9A.3   Road Sector Diesel Consumption in the Dominican Republic, 1998–2008

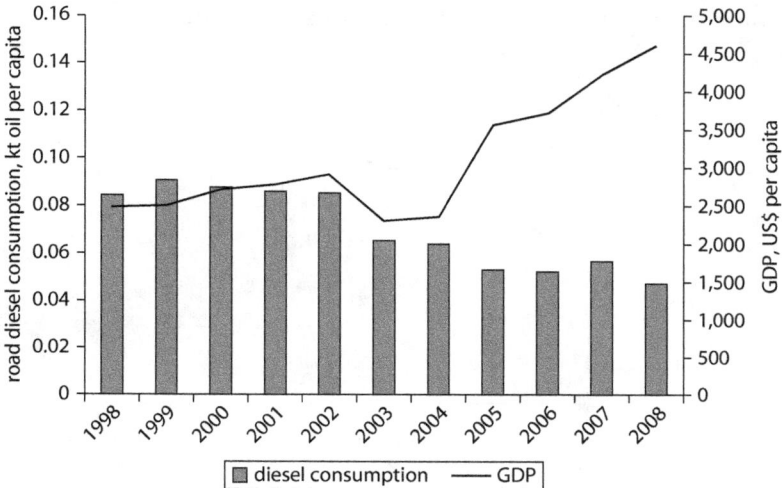

Source: World Bank, World Development Indicators.

Figure 9A.4   Road Sector Gasoline Consumption in the Dominican Republic, 1998–2008

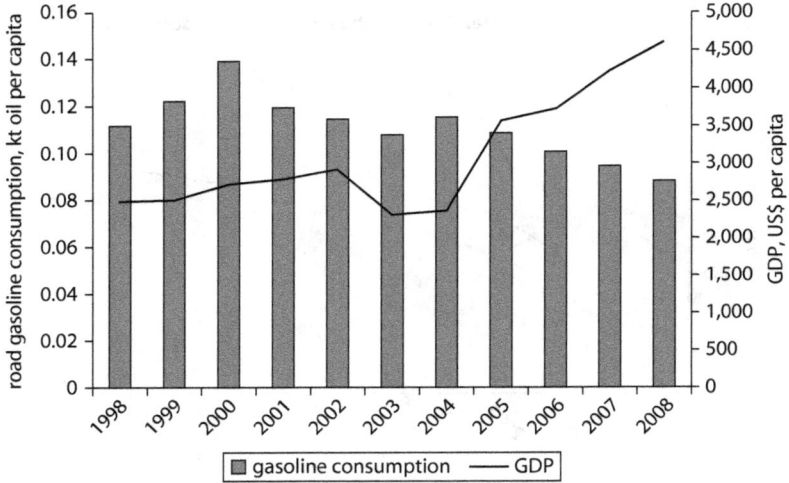

Source: World Bank, World Development Indicators.

Electricity Price and Power Consumption in the Dominican Republic

Figure 9A.5   Electricity Price in the Dominican Republic, 2002–10

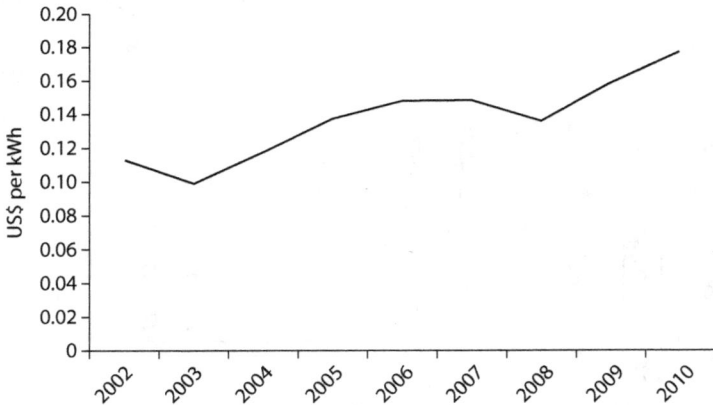

Source: SIE 2011.
Note: kWh = kilowatt-hour.

Figure 9A.6   Power Consumption Per Capita in the Dominican Republic, 1998–2008

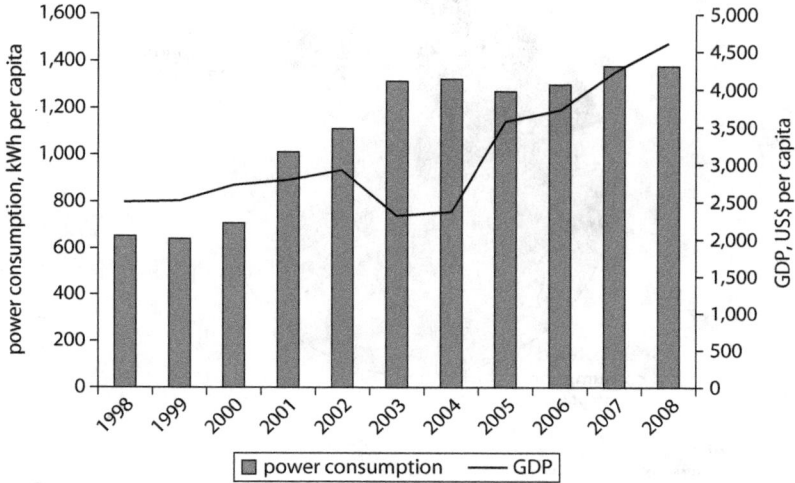

Source: World Bank, World Development Indicators.
Note: kWh = kilowatt-hour.

Poverty Impact Evidence from Household Surveys in the Dominican Republic

Figure 9A.7   Electricity Block Tariffs in the Dominican Republic, 2010

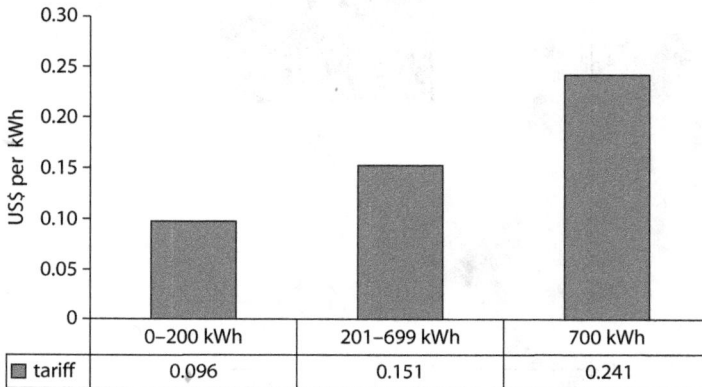

	0–200 kWh	201–699 kWh	700 kWh
tariff	0.096	0.151	0.241

Source: SIE 2011.
Note: kWh = kilowatt-hour.

Figure 9A.8   Metered Electricity Consumption in the Dominican Republic, 2010

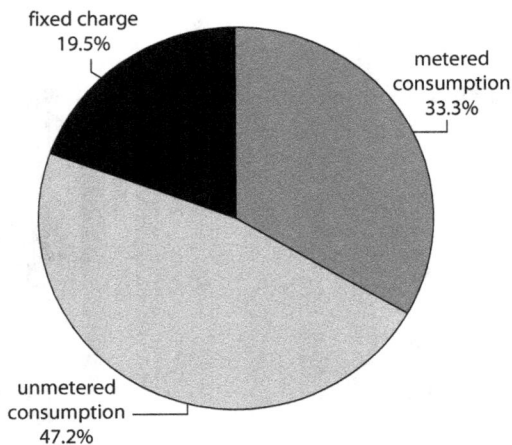

Source: SIE 2011.
Note: "Fixed charge" refers to one-time connection charges only.

Figure 9A.9   Average Household Expenditure on Energy in the Dominican Republic, by Energy Type, 2007

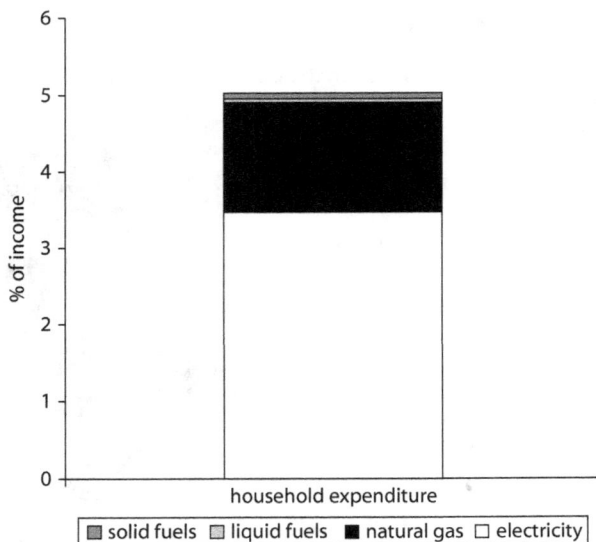

Source: ONE 2007.

Figure 9A.10   Subsidies and Social Programs for the Poor in the Dominican Republic, 2006–09

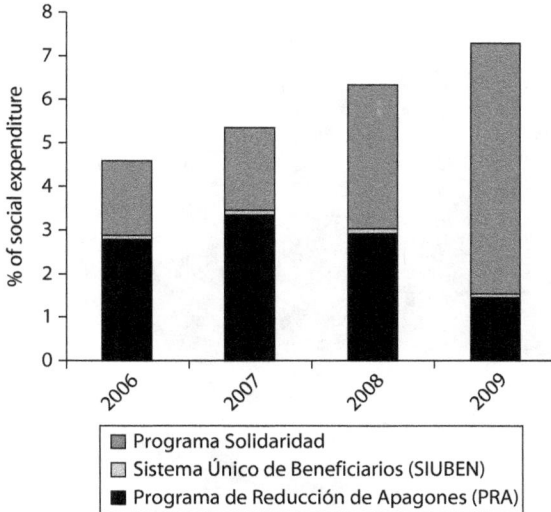

Source: Ministerio de Economía, Planificación y Desarrollo.

References

IDB (Inter-American Development Bank). 2009. "Dominican Republic Gets $500 Million Loan from the IDB." Press release, October 21.

IMF (International Monetary Fund). 2011. "Dominican Republic: Fourth Review under the Stand-By Arrangement and Request for Waiver of Nonobservance of Performance Criterion." Country Report 11/177, IMF, Washington, DC.

ONE (Oficina Nacional de Estadística). 2007. "Encuesta Nacional de Ingresos y Gastos de los Hogares (ENIGH)." Household survey, ONE, Santo Domingo.

SEIC (Secretaría de Estado de Industria y Comercio). 2011. Government website. http://www.seic.gov.do. Accessed June–July 2011.

SIE (Superintendencia de Electricidad). 2011. Government website. http://www.sie.gov.do. Accessed June–July 2011.

World Bank. 2009. "Dominican Republic: World Bank Approves US$300 Million to Support Social Sectors and Public Finance." Press release, November 17.

CHAPTER 10

Peru

Incentives to Energy Subsidy Reforms

Peru is well endowed with hydroelectric and natural gas sources of energy. Peru's hydropower potential is considered to amount to at least 60,000 megawatts (producing about 400,000 gigawatt-hours annually). It has about 11.8 trillion cubic feet of proven reserves of natural gas. Historically, Peru has relied mostly on hydropower, but after the entry into production in 2004 of natural gas from the huge Camisea field, electricity generation capacity became evenly divided between thermal and hydropower. Most electrical energy is still produced from hydropower, with thermal plants mainly used during daily peak-load periods and in seasons when hydropower output is lower than average.

Peru is trying to further reduce its dependence on hydroelectricity because capacity varies with water levels and rainfall. The thermal sources that gained additional use are diesel, coal, and oil, but a shift toward liquefied natural gas (LNG) has been the most recent energy strategy for Peru. The Camisea natural gas project is likely to have a major impact on both tariff rates and fuel use, as natural gas consumption is expected to rise significantly. An LNG plant capable of processing 4.4 million tons a year is expected to also enable exportation of natural gas to neighboring countries.

Fossil Fuels

In Peru, fuel prices were already quite high even before the oil price rally because of the high excise tax, but the government has tried to limit the impact on inflation through a temporary excise tax cut and a price stabilization scheme funded by the Treasury. The Fuel Price Stabilization Fund (FEPC), guaranteed by the state, was created in 2004 with the objective of reducing the price volatility of the international fuel market. Firms that keep prices within an official wholesale price band set by the authorities receive compensation for the gap between the band and an import parity reference price formula (see figure 10.1). The authorities also reduced specific fuel excises to limit the pass-through to domestic retail prices.

The fiscal cost of the fuel price subsidies reached a peak in July 2008, when the cost for the year was projected at 1.4 percent of gross domestic product (GDP). Of this cost, 1.2 percent of GDP corresponded to the pure fuel price subsidy, and 0.2 percent to ad valorem tax revenue forgone due to lower retail prices. In addition, the annual fiscal cost of the fuel excise tax cuts during 2004–08 was estimated at 0.5 percent of GDP, bringing the total 2008 cost of fuel price stabilization policies to 1.9 percent of GDP. The average fuel price increase required to eliminate the subsidy peaked at 45 percent. Table 10.1 presents the companies' contributions and the government's compensations during 2004–09.

Figure 10.1   Functioning of the Fuel Price Stabilization Fund in Peru

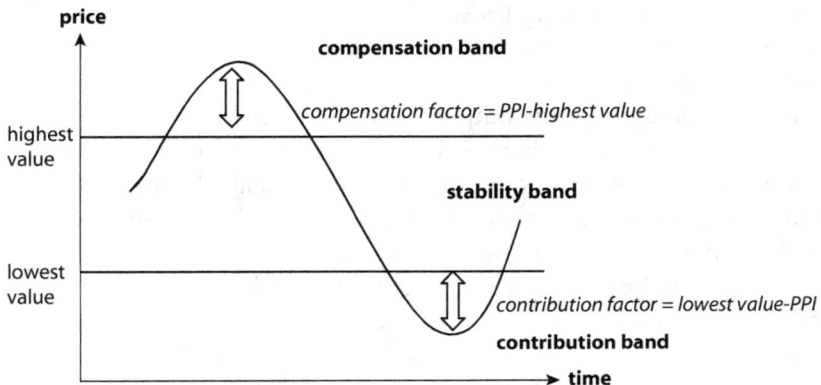

Note: PPI (parity import price) = PR1 (reference import price) + average wholesale sale margin.

Table 10.1   Contributions and Compensations of the FEPC in Peru, 2004–09
US$

Year	Balance previous year	Contributions by companies	Compensation by government	Transfers by government	Total
2004/05	0	25,040,946	−91,226,129	0	−65,926,466
2006	−65,986,466	50,154,686	−37,385,968	0	12,741,172
2007	12,765,957	7,315,741	−381,142,513	69,166,343	−298,507,465
2008	−298,454,260	229,041,536	−1,172,915,576	782,672,011	−460,138,336
2009 (Q1)	−459,719,943	127,407,085	0	364,033,494	31,234,074

Source: Ministry of Energy and Mining of Peru 2009.
Note: In 2007, the government declared an LPG market emergency, authorizing compensations to importers of US$6,552,602 through the FEPC. FEPC = Fuel Price Stabilization Fund.

Electricity

The electricity subsidy model applied in Peru is one that seeks to determine assistance based on economic cost of service and household amount of energy use per day. Because the costs of supplying electricity in rural areas are higher than those for urban areas, the tariffs would also need to be higher in rural areas to account for expanding the electrical systems. The level of investment is estimated to be two to five times greater for these areas (Revolo 2009).

Peru has one of the lowest rural electrification rates in Latin America. The government of Peru tried to extend access to basic infrastructure services, including electricity, to the dispersed population living in these areas. Plans and targets have been in place for rural electrification since the early 1970s, but by 2005, only 39 percent of rural households had electricity service (World Bank 2010a).

Currently, three types of subsidies have been adopted to provide equal electricity access to low-income households (Revolo 2009; World Bank 2010a):

- *Funding from the government and international donors*—an investment subsidy for the capital costs of new distribution, which typically includes isolated mini-grid projects
- *Internal tariff subsidies*—a system of cross-subsidies that includes two components aimed at (a) reducing the prices of generation in isolated systems, and (b) compensating for the differences in distribution costs between urban and rural areas
- *The Electricity Social Compensation Fund (Fondo de Compensación Social Eléctrica; FOSE)*—a consumption cross-subsidy, taken from urban

high-electricity consumers (those using more than 100 kilowatt-hours [kWh]) and given to rural customers (those using less than 100 kWh), as further explained in the "Reform Efforts" section below.

Natural Gas

Although not a subsidy in the technical sense, natural gas prices for power generation in Peru are among the lowest in the world, largely as a consequence of the pricing policy that set a cap on the Camisea wellhead price for power generation. The price for natural gas was introduced by the government of Peru to promote the use of newly available natural gas after the development of the country's Camisea gas fields. The low price has made natural gas the preferred fuel for power generation over the past 10 years. However, in carrying out an economic analysis of hydropower generation, the World Bank found that hydropower, not natural gas, is the least costly option for power generation if valued on an economic basis. According to the study, the natural gas prices are distortionary because they make it impossible for hydropower projects to compete in energy generation (World Bank 2009).

The resulting economic prices of natural gas for power generation in the Lima area are presented in table 10.2 as a function of oil price scenarios. Oil and natural gas have natural links in their production and consumption (as competing fuels) and hence, irrespective of short-term volatilities in the prices of either fuel, they keep a strong long-term relationship.

Table 10.2   Economic Value of Natural Gas

Oil price (US$ per bbl)	Economic gas price (LNG export netback) (US$ per MMBtu)
37	2.14 (2009 gas price)
75	4.4
100	5.9
125	7.3

Source: World Bank 2009.
Note: LNG = liquefied natural gas; bbl = barrel; MMBtu = million metric British thermal units.

Reform Efforts

Fossil Fuels

In response to the sharp increase in oil prices, the authorities increased the official FEPC price band in January, March, June, and August 2008 by a cumulative 28 percent. They also made adjustments to the import parity price formula to ensure that it did not overstate true import costs and began unwinding the specific excise tax cuts in November 2008, when these taxes were more than doubled.

The International Monetary Fund assessed that, at 2010 prices, the potential liability of the FEPC would amount to 0.5 percent of GDP in 2010. The authorities were considering moving to a rule-based adjustment of the price bands (to ensure that the FEPC is self-financed) and to an automatic settlement of liabilities. A gradual but frequent adjustment of the reference price can help minimize disruptive effects on other prices and avoid the buildup of large price differentials and debt obligations. This adjustment could be facilitated by placing the oversight of the FEPC under a technical regulatory agency. The changes in the FEPC framework could be accompanied by a mechanism to minimize the effects on the more vulnerable segments of the population (IMF 2010).

Electricity

Reform of the power sector started with the passage of the Electric Concessions Law in 1992 with its regulations[1] and Supreme Decree[2] in 1993 that established a new legal framework that provided for (a) the restructuring in 1992 and 1993 of the main vertically integrated power suppliers into separate generation, transmission, and distribution functions; and (b) open access to the transmission and distribution networks by generators, power traders, and large power users.

The new legal framework also provided for the creation of a sector regulator, the Energy and Mining Investment Supervisory Board (*Organismo Supervisor de la Inversión en Energía y Minería*; OSINERGMIN). The role of the state was limited to sector policy and general regulations, the granting of concessions, and basic sector planning. Privatization in the power sector started in 1994 and progressed successfully until 1997, when major sector assets comprising 70 percent of generation capacity, 100 percent of transmission capacity, and 45 percent of the distribution market were transferred from public to private ownership, management, and operation.

The Peruvian electricity tariff scheme is designed on the basis of full cost recovery in each of the three segments: generation, transmission, and distribution systems. The generation regulated energy tariff is determined by OSINERGMIN every year according to the expected evolution of demand and generation supply capacity, fuel prices, competitive generation auction prices, and other economic parameters (such as price indexes and inflation). The System Economic Operation Committee (COES) carries out real-time dispatch of generation supply by following a cost-based, merit-order procedure independent of bilateral contracts or the results of energy auctions. Hourly transactions between generators, distribution companies, and large users in the wholesale market are done at the marginal energy price.

These reforms helped Peru expand its electrification from 57 percent in 1993 to 75 percent in 2002 (Cherni and Preston 2007). Nevertheless, the electrification efforts left out the rural areas, which were not given serious consideration until the law that established FOSE in November 2001, giving rise to cross-subsidies and the 2002 Rural Electrification Law.

Since July 2004, the level of subsidy for FOSE has consisted of tariff reductions for monthly consumption up to 30 kWh—set at 25 percent for urban users supplied by the interconnected system and 62.5 percent for rural users supplied by isolated systems. For consumption of 31–100 kWh, the reduction is gradual, from a maximum of 31.25 percent for rural users supplied by isolated systems to a minimum of 7.5 percent for urban users supplied by the interconnected system. Consumers who use more than 100 kWh per month pay a cross-subsidy in proportion to their energy consumption above 100 kWh per month to finance the FOSE discount.[3]

The rationale for FOSE is regional equalization of tariffs for those at the lower levels of consumption, with the general objective of reducing the differential between the high tariffs of the outlying provinces and the lowest tariff in Lima. (For that reason, the tariffs found in figures 10A.1 and 10A.4 should not be taken as representative values for the whole country but rather as indications of an overall trend.) Sixty percent of all customers benefit from FOSE. The total FOSE transfer in 2004 was US$18 million (relative to a total consumer bill of US$600 million). The recovery mechanism increases bills to all consumers with consumption greater than 100 kWh per month by 2.5–3 percent. The electricity tariff faced by low-income rural households is a complex, nonlinear function of monthly consumption—a consequence of the approach to rate making adopted by OSINERG (later renamed OSINERGMIN) and the cross-subsidy system adopted to finance the lifeline tariff rate of FOSE (World Bank 2010b).

In addition to FOSE, Law No. 28832 of July 2006 (to "Ensure the Efficient Development of Electricity Generation") introduced a new provision whereby electricity users being served through the national interconnected system provide, by way of an increase in their electricity tariffs, financial support to users connected to isolated systems to reduce the generation costs of such systems.

In August 2006, in cooperation with the World Bank, an electrification project was launched to assist local distribution companies in reaching rural populations with well-targeted subsidies, aiming at financing projects that would be financially sustainable after receiving a subsidy of a substantial part of the capital costs (World Bank 2010a).

Poverty Alleviation Measures

Social Safety Nets

Unlike other countries where governments use social data to identify household income levels for subsidy purposes, in Peru the households are subsidized if they use less than 100 kWh per month under the assumption that low-income families use less electricity (see figure 10A.6). Because of low access and metering, the richest quintile consumes more electricity as a percentage of income. The increasing block tariff is regressive, with the top two quintiles benefiting from almost half of the subsidy (see figure 10A.8).

In February 2005, the Juntos cash transfer program was launched separately by the Alejandro Toledo administration. The goals of the program are to reduce poverty by providing households with cash transfers in the short run and to improve access to education and health services in the long run. The selection of the beneficiary households occurs in three stages: selection of eligible districts, selection of eligible households within the eligible districts, and finally a community-level validation that finalizes the actual beneficiary list. Participating districts were selected on the basis of criteria such as exposure to violence, poverty level, poverty gap, level of child malnutrition, presence of extreme income poverty, and other factors.

A recent impact analysis carried out by the World Bank suggests that Juntos is improving a number of key welfare indicators of program beneficiaries. Specifically, Juntos has a moderate impact in reducing poverty and increasing monetary measures of both income and consumption. In addition, and similar to evidence from other countries, the program increases the utilization of health services for both children and women, and it improves the nutritional intake of program households. In education, the analysis shows that as in other conditional cash transfer contexts where primary school attendance is high, Juntos has impacts mainly at transition points, ensuring that children enter and finish primary school (Perova and Renos 2009).

Evidence from Household Surveys

In order to assess the demand and use of electricity in rural areas of Peru and measure the performance of the current subsidy schemes, a household survey was conducted in 2005 by local authorities together with the World Bank. The special report on energy and poverty that analyzed the data from that survey found that although FOSE would significantly reduce the energy charge to rural consumers (by 50 percent if supplied by the interconnected system, by 62.5 percent in the case of isolated systems), it would at the same time marginalize end users who consume less than 15 kWh per month and benefit them much less in terms of the value of the subsidy (in nuevos soles) than the average consumer who uses 25–35 kWh per month (World Bank 2010b). This disparity occurs because, at low consumption levels, the fixed charges (connection fees) dominate the bill. Further, the report found that households in the lowest quintile capture only 7.7 percent of the total FOSE subsidy received by all rural households, although that quintile constitutes 20 percent of all households. At the same time, the highest quintile captures 32.6 percent of the benefit. The report suggests that those who consume small amounts of electricity pay relatively high prices per kWh, the FOSE mechanism notwithstanding, and concludes that the targeting performance of FOSE is poor (World Bank 2010b).

Key Lessons Learned

Peruvian electricity sector reforms are generally regarded as successful. The electricity subsidies are not typically viewed as wasteful or perverse because they have been introduced by the government to make electricity for the rural poor as affordable as for urban dwellers. Moreover, the fact that consumers benefit from this subsidy if they use less than 100 kWh per month encourages users to keep energy use at a thrifty level.

Similar to Chile, the power sector in Peru is unbundled into separate generation, transmission, distribution, and retail markets. Creation of OSINERG (later named OSINERGMIN) as an independent regulator for determining and supervising tariffs for regulated customers is also part of Peru's successful energy sector reform. The use of cross-subsidies in the case of rural electricity consumption and isolated mini grids, which typically carry higher operating costs, is well administered and could serve as country experience for those looking to increase electrification rates and make electricity affordable for the poor living in remote areas.

A recent study conducted by the World Bank comes to the conclusion that there is room for improvement in the FOSE scheme (World Bank 2010b). In that context, the report suggests that improvements in targeting performance could be achieved by further lowering the FOSE cap. If the 50 percent discount were limited to 15 kWh per month and phased out at 25 kWh per month, the share of benefits going to the lowest quintile would be 19 percent, while the richest would receive less than 10 percent.

Annex 10.1 Peru Case Study Figures

INCOME LEVEL: Upper-middle income
REGION: Latin America and the Caribbean
ENERGY NET IMPORTER/EXPORTER: Net importer (potential net exporter)
SUBSIDIES: Electricity
PHASING OUT SUBSIDIES: Ongoing

Fuel Prices and Road Sector Consumption in Peru

Figure 10A.1   Domestic Retail Fuel Prices in Peru, 2002–10

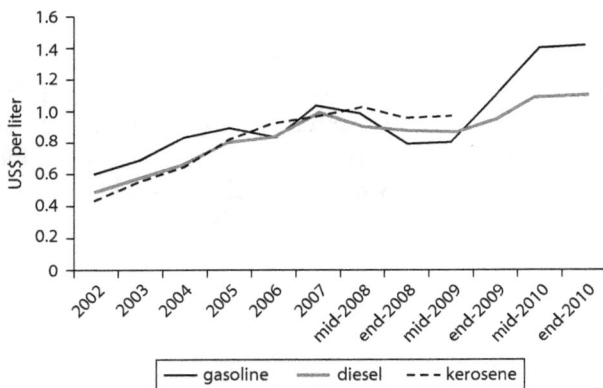

Source: INEI n.d.

Figure 10A.2   Road Sector Diesel Consumption in Peru, 1998–2008

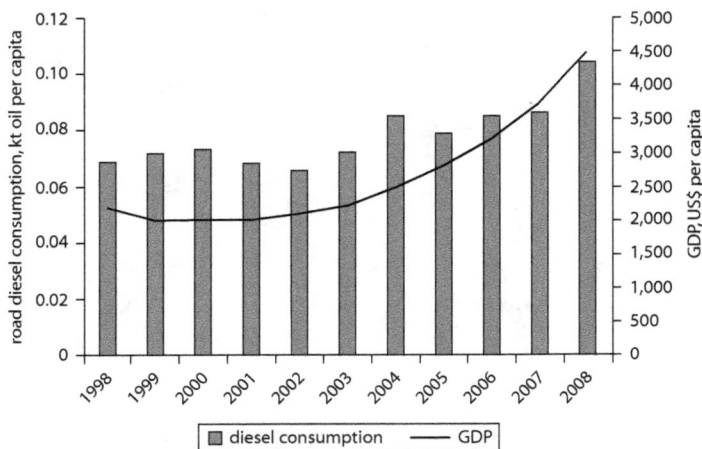

Source: World Bank, World Development Indicators.

Figure 10A.3   Road Sector Gasoline Consumption in Peru, 1998–2008

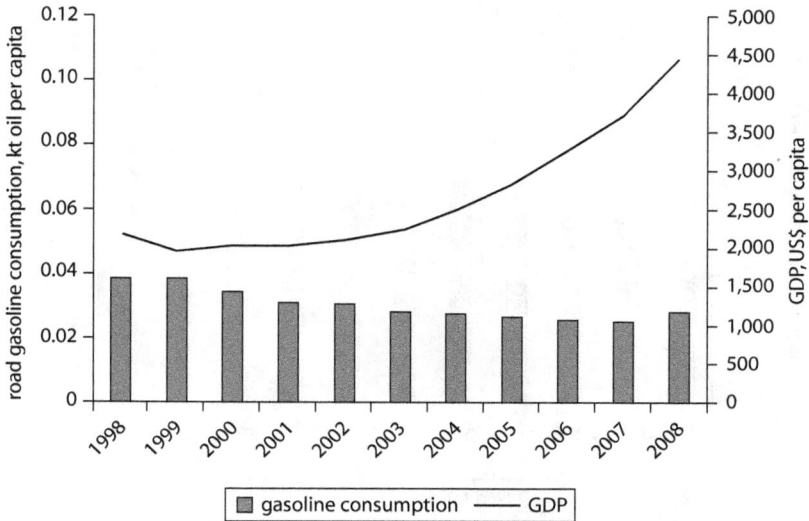

Source: World Bank, World Development Indicators.

Electricity Price and Power Consumption in Peru

Figure 10A.4   Electricity Price in Peru, 1998–2010

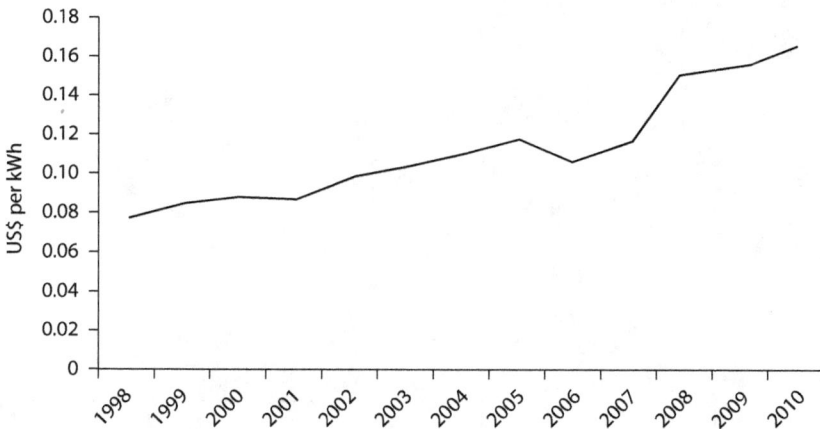

Source: INEI n.d.
Note: kWh = kilowatt-hour.

Figure 10A.5   Power Consumption Per Capita in Peru, 1998–2008

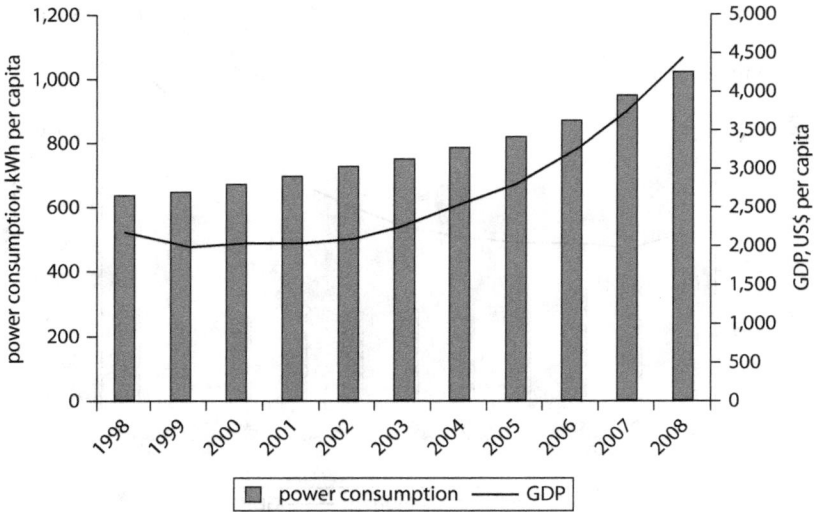

Source: World Bank, World Development Indicators.
Note: kWh = kilowatt-hour.

Poverty Impact Evidence from Household Surveys in Peru

Figure 10A.6   Electricity Block Tariffs in Peru, 2002–11

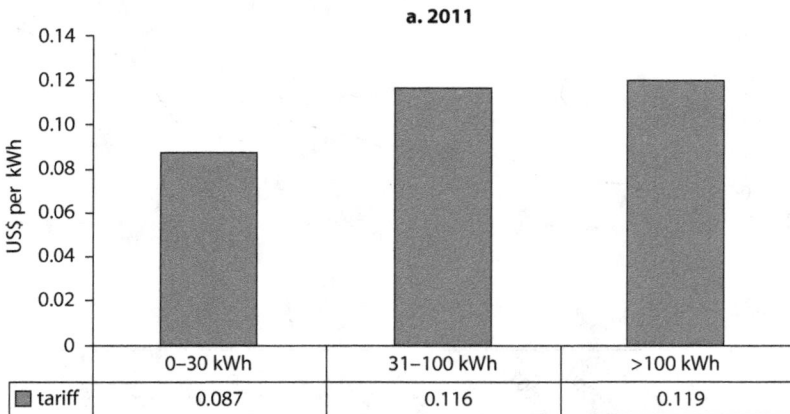

	0–30 kWh	31–100 kWh	>100 kWh
tariff	0.087	0.116	0.119

(continued next page)

Figure 10A.6   *(continued)*

b. 2002–11

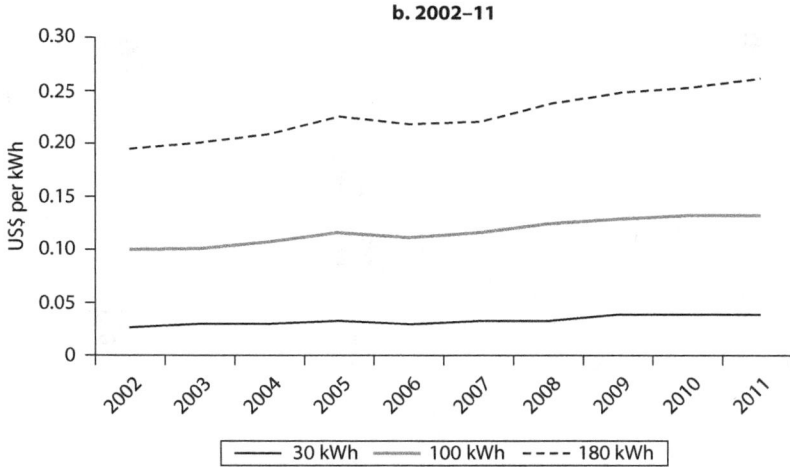

Source: INEI n.d.
Note: kWh = kilowatt-hour.

Figure 10A.7   Electricity Expenditure in Peru, by Income Quintile, 2003

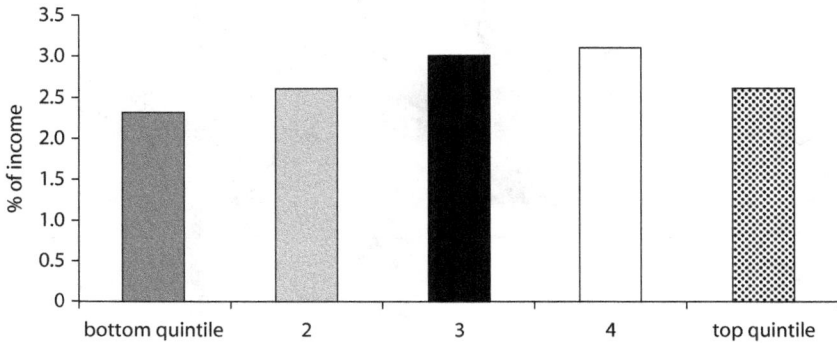

Source: INEI 2005.

Figure 10A.8   Benefit Incidence of Electricity Subsidies in Peru, by Income Quintile, 2003

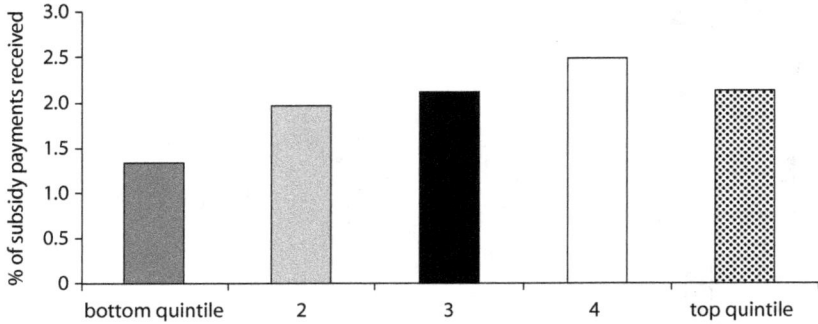

Source: INEI 2005.

Figure 10A.9   Welfare Impact of Fossil Fuel Subsidy Removal in Peru, 2003

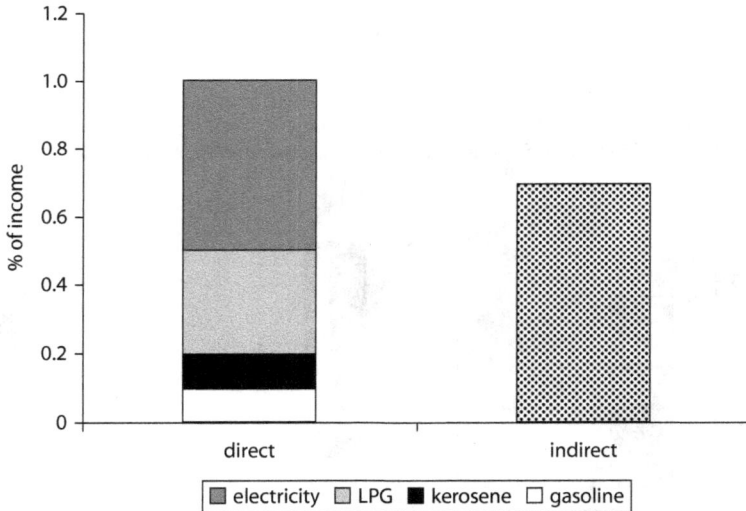

Source: Del Granado, Coady, and Gillingham 2010.
Note: LPG = liquefied petroleum gas. "Indirect" impacts include higher prices for non-energy goods and services consumed by households to the extent that higher fuel costs increase production costs and thus consumer prices.

Notes

1. Law No. 25844.
2. No. 009-93-EM.
3. It should also be noted that rural tariffs vary by location, based on the tariff calculated by OSINERG (later named OSINERGMIN) for the areas of each distribution company. The price paid per kWh (from the 2005 household survey) varied from a low of S/.0.47 per kWh in the South Coast region to a high of S/.0.83 per kWh in the Andean South region.

References

Cherni, Judith, and Felix Preston. 2007. "Rural Electrification under Liberal Reforms: The Case of Peru." *Journal of Cleaner Production* 15 (2): 143–52.

Del Granado, J., D. Coady, and R. Gillingham. 2010. "The Unequal Benefits of Fuel Subsidies: A Review of Evidence for Developing Countries." Working Paper 10/202, International Monetary Fund, Washington, DC.

IMF (International Monetary Fund). 2010. "Peru: Staff Report for the 2010 Article IV Consultation." Country Report 10/98, IMF, Washington, DC.

INEI (Instituto Nacional de Estadística e Informática) (database). 2005. "Encuesta de consumo de energía a hogares en el ámbito rural 2005." Household survey data, INEI, Lima.

———— (database). n.d. INEI, Lima. http://www.inei.gob.pe. Accessed June–July 2011.

Ministry of Energy and Mining. 2009. "Informe situación del fondo para la estabilización de precios de los combustibles." Report, March 31. Ministry of Energy and Mining, Lima.

Perova, Elizaveta, and Vakis Renos. 2009. "Welfare Impacts of the 'Juntos' Program in Peru: Evidence from a Non-Experimental Evaluation." Impact Evaluation Paper, World Bank, Washington, DC.

Revolo, Miguel. 2009. "Mechanism of Subsidies Applied in Peru." Presentation at Africa Electrification Initiative Workshop, Maputo, Mozambique, June 11.

World Bank. 2009. "Peru: Overcoming Barriers to Hydropower." Energy Sector Management Assistance Program (ESMAP) study, World Bank, Washington, DC.

————. 2010a. "Addressing the Electricity Access Gap." Background paper for the World Bank Group Energy Sector Strategy, World Bank, Washington, DC.

————. 2010b. "Peru: National Survey of Rural Household Energy Use." Special Report 007/10, World Bank, Washington DC.

CHAPTER 11

Turkey

Incentives to Energy Subsidy Reforms

Fuel Pricing

Turkish gasoline and diesel prices are currently among the highest in the Organisation for Economic Co-operation and Development (OECD) countries, owing to the relatively high excise taxes that are reflected at the level of retail prices. A recent study, using a structural vector autoregression methodology and monthly data, argues that there is an asymmetry in how gasoline prices in Turkey respond to crude oil price changes. When crude oil prices increase, this is reflected in higher tariffs, but when crude oil prices decrease, there is no response. An example of how different tax components determine the final gasoline price shows that the refinery price made up only 15.6 percent of the retail price. The transportation cost, distributor's share, and supply station's share combined amounted to 14.2 percent, while the total tax (special consumption and value added taxes) made up 70.2 percent of the retail price (Alper and Torul 2009).

Electricity

The impetus behind electricity sector reforms, among other factors, has been Turkey's long-standing ambition to join the European Union (EU).

A reform program to harmonize the Turkish electricity sector with the EU energy acquis became especially prominent after 2001 with Turkey's announcement of the Electricity Market Law (EML). As part of the gas and electricity market reform, Turkey has moved toward a fully cost-reflective tariff structure (Bagdadioglu, Basaran, and Price 2008).

Hidden costs coming from underpricing, lack of collection, and unaccounted losses were substantially reduced as the results of such reforms, from the peak of 2.1 percent of gross domestic product (GDP) in 2001 to 0.6 percent of GDP in 2003 (see figure 11A.1). Natural gas hidden costs have remained a challenge because natural gas is mainly used for heating and has been subsidized to protect consumers.

One of the persistent problems with Turkey's electricity tariff system has been the use of cross-subsidization, whereby households would be charged below-cost prices at the expense of the industrial sectors (Cetin and Oguz 2007). Cross-subsidies were also applied between different regions.

Reform Efforts

Fossil Fuels

The Automatic Pricing Mechanism, which operated between July 1998 and the end of 2004, set a ceiling on the prices of almost all oil products in Turkey. At the beginning of 2005, the government decided to remove the price caps, which led to an increase in pretax prices. Since then, oil prices have been set by the market.

Turkey is also providing financial assistance through transfer payments from the Turkish Treasury to the state-owned Turkish Hard Coal Enterprises. The Ministry of Energy and Natural Resources distributes coal for heating purposes to assist poor families.

Electricity

In 1993, the Turkish Electricity Authority, which used to have a monopoly in the energy sector, was unbundled into the Turkish Electricity Generation Transmission Company (responsible for both generation and transmission activities) and the Turkish Electricity Distribution Company (TEDAS) (responsible for distribution and retail sale activities).

The Energy Market Regulatory Authority (EMRA) was established in 2001 to issue licenses for electricity generation transmission and distribution and to set electricity tariffs. However, political friction has occurred in the past between EMRA and the Ministry of Energy and Natural

Resources. When EMRA tried to introduce a cost-based regional pricing scheme in 2003, some government members opposed it and hence the plan was dropped (Cetin and Oguz 2007).

Although wholesale prices for the gas and electricity markets are already cost-based, the retail prices remain regulated by means of a uniform national retail tariff, which is approved by EMRA. Hence, the retail tariff does not reflect the differences in costs across the distribution regions.

In 2001, the EML was passed as a fundamental approach to easing the burden of the power sector on the public budget. The law provided for the unbundling of the state-owned electricity assets; opened the market above a certain level of electricity consumption (the threshold to gradually decline); and allowed third-party access to the grid. Although corporatized with separate accounts, these entities remained under government control in terms of decision making, and they had little managerial autonomy. The EML also established EMRA as an independent and financially autonomous regulator of power, gas, petroleum, and liquefied petroleum gas, supervised by the Energy Market Regulatory Board. EMRA issues licenses for electricity generation transmission and distribution and sets electricity tariffs (Cetin and Oguz 2007).

In 2004, a strategy paper was approved by the Higher Planning Council. It outlined the steps for further liberalization of the electricity sector. The strategy paper also provided the basis for determining the revenue requirements of the regional distribution companies ex ante.

In April 2006, TEDAS, with its 20 regional distribution companies, was transferred to the Privatization Administration, which decided to commence the privatization of the electricity distribution sector. As a result, the share of private ownership in the distribution sector changed dramatically—from a minor portion in 2008 toward 100 percent in 2011. Revenues from privatization represent a significant source of budget revenues. Four distribution companies sold in 2008 brought about US$2.4 billion into the budget. Seven other companies were sold for about US$2.7 billion in 2009. The remaining companies were expected to be sold in 2010. The privatization was expected to help improve the situation with bill collection and thus to increase the payments to suppliers. The results of the 2008 round of privatization indicate that four privatized distribution companies had almost 100 percent bill collection rates and paid their bills in full to generators and electricity suppliers (World Bank 2010).

Next the government plans to extend the privatization into the generation sector. The major generation privatization program was announced in March 2010. The share of the private sector in generation, which was 50 percent in 2008, is expected to increase to 75-80 percent by 2015. The privatization should involve all of EÜAŞ's (the state-owned electricity generation company) thermal capacity and about 50 percent of its hydro capacity. Most of these assets would be offered for sale in nine portfolio groups, and 3,074 megawatts in four priority plants would be sold independently. The portfolios were formed in ways that would attract a wide variety of investors (World Bank 2010).

A competitive wholesale electricity market went into operation in 2006. In April 2009, EMRA issued new balancing and settlement regulations to improve the functioning of the wholesale electricity market. In December 2009, the market moved from monthly settlement to hourly settlement. By 2010, approximately 400 private companies were trading power in this market, through which on average about 30 percent of total electricity supply is dispatched. In July 2008, the EML of 2001 was substantially amended to promote energy security.[1]

Between 2002 and 2007, retail prices for electricity remained constant in spite of a significant increase in generation costs due to high fuel input prices. Constant prices, along with problems with network losses and bill collection, had a negative effect on the financial viability of the sector, limited the availability of funding for new investments, and did not send the right price signals to consumers. The new cost-based pricing mechanism was put in place in 2008. Under the new regime, tariffs are adjusted quarterly to take into account the increases in costs incurred by utilities, including increases in input prices, inflation, and exchange rates. The transition to the new system involved three large tariff increases (in January, July, and October 2008) that raised the average retail tariff by about 15 percent, 24 percent, and 9 percent, respectively (see figure 11A.5). The new cost-based mechanism includes automatic price adjustments for future cost increases. The implementation of the new pricing mechanism has already resulted in the payments of current bills and a significant reduction of arrears to private generators (World Bank 2010).

From 2013 on, tariffs for electricity distribution will provide incentives for privatized operators to reduce the proportion of electricity that is lost or stolen each year—a loss that stood at about 14 percent in 2009.

Poverty Alleviation Measures

Electricity is mainly used in Turkish households for lighting, power, and air conditioning, with little used for heating, a need mainly met by burning oil, coal, or natural gas (in larger cities).

Turkey has applied a "national" system of residential electricity tariffs, with a small discount (0.65 percent) for priority provinces—mainly, the more rural provinces in the south and east of the country. This flat national rate does not mirror the very high level of distributional losses and regional differences in distribution losses. The priority discount status reflects lower average income in the selected provinces, and one reason to delay the introduction of tariffs that reflect differential regional losses is likely to have been on distributional grounds. The level of residential tariffs in Turkey has been almost equal to those charged to industrial consumers, that is, a much lower proportion of industrial charges than the average in EU, other OECD, and Southeast European countries. (The average ratio is 1.7 percent in OECD countries.) Turkey's flat rate for each unit of electricity consumed represents an interesting compromise between a cost-reflective tariff and an affordability-focused one.

Evidence from Household Surveys

To make tariffs cost-reflective, considerable change in the tariffs is necessary. Bagdadioglu, Basaran, and Price (2008) modeled the effects on households that may arise from reforms:

- *Scenario 1*, simulating changes in regional tariffs, demonstrates the effect of charging provinces for the losses they incur. It is found that impacts are more pronounced for the poorest decile, leading to an increased expenditure of up to 12.5 percent for electricity (see figure 11A.8). This is to be explained by the fact that the losses would be recovered from consumers who currently have no electricity expenditure but nevertheless use power.

- *Scenario 2*, simulating a change in the level of revenue collected from residential users (a price rise of 14 percent on average), would cause the poorest decile to spend around 5 percent more for electricity, while the change for the richer decile would be less pronounced.

- *Scenario 3*, simulating the introduction of a revenue-neutral standing charge (of 10 percent of the average bill), would—again, even though to a lesser extent—benefit those who consume more electricity and penalize users of small quantities, as figure 11A.8 shows (Bagdadioglu, Basaran, and Price 2008).

Increases in the electricity tariff that took place in 2008 under the new cost-based pricing mechanism coincided with the economic crisis and a decline in the country's economic growth. As a result, electricity expenditures increased for all groups of customers (see figure 11A.9). The share of electricity expenditures increased from an average of 2.9 percent of the households' disposable income in 2007 to 3.5 percent in 2008. The average budget share for electricity in the lowest quintile is still below the 10 percent affordability benchmark. In 2008, Turkey had the eighth-lowest average household share of electricity expenditures among European and Central Asian countries.

The low-income households experienced a larger welfare loss from the price increases than higher-income households. The analysis conducted by the World Bank estimates that the welfare loss from the electricity price increases of 2008 was 2.16 percent of disposable income of the bottom quintile of the households while the loss of the top quintile was 0.75 percent. The crisis impact survey conducted by the World Bank, the United Nations Children's Fund, and the Turkish think tank TEPAV (Economic Policy Research Foundation of Turkey) in five urban areas indicates that more than 50 percent of the poor reported problems with payments for electricity and that 9 percent of electricity subscribers experienced disconnections that followed nonpayments. No significant increases in electricity prices were expected in the future and thus there should be only a modest welfare impact (World Bank 2010).

Social Safety Nets

Turkey's social safety net system is called the Social Risk Mitigation Program (SRMP) and is administered by the General Directorate of Social Assistance and Solidarity (GDSAS), which provides assistance in two forms: social insurance and social assistance.

The system as a whole includes institutional capacity development, a component of GDSAS, which is the largest institution dealing with social assistance and social transfers. The local initiatives component provides some support for poor families to set up their small businesses or to support their young children with training programs. The rapid response component provides in-kind transfers or cash transfers, but one time only in the form of heating and food support.

Among the core services of SRMP are conditional cash transfers (CCTs), which are primarily provided to improve educational and health standards. The CCTs were introduced after two economic crises and an earthquake in the Marmara region in 2001. The main channels for cash transfers are post offices as well as bank accounts that are set up for distribution of the transfers. The cash transfer program replaces another program that previously used means testing and ran into proper identification problems due to Turkey's large informal sector. Turkey's CCT program now uses the scoring system of proxy means testing, based on observable characteristics such as housing location, living quality, ownership of different goods, and so forth. Cash transfer program benefits are adjusted according to Turkey's annual change in the consumer price index to account for inflation. According to reports, the program is effective in reaching the poor and others for whom the assistance is intended, but as with most safety net programs, better coordination and payment delay reductions are needed (Gokalp 2007). Poverty levels are also monitored using regular surveys to measure progress of the program.

The program targets poor families with children who are ages zero to six or in primary or secondary school, and pregnant mothers (World Bank n.d.). The conditionality asked of these families is to send their children to school and take them to health centers on a regular basis. Officials reported that the program was quite successful in terms of school attendance while immunization increased by up to 14 percent (Fiszbein and Schady 2009). School enrollment rates among girls were lower than among boys, and the CCTs have helped to reduce this gender gap.

It is likely that the target of reaching the poorest 6 percent of the population has not been reached yet, despite the good targeting performance of the scheme. A quantitative assessment concluded that the CCT program was well targeted to the poorest and affirmed that the income distribution of CCT education and health beneficiaries was highly progressive (World Bank 2008). High percentages of all beneficiaries belonged to poorer income groups, particularly the health beneficiaries. Fifty percent of all health-beneficiary households and over 30 percent of all education-beneficiary households were among the poorest 10 percent of all households in the income distribution at the national level. Indeed, no education beneficiaries or health beneficiaries belonged to the richer 30 percent and 40 percent of all households, respectively. At the same time, the evaluation results suggest that although the CCT program effectively reached the poorest, a considerable number of non-beneficiary-applicant households were also among the poorest but excluded from the program (World Bank 2008).[2]

Key Lessons Learned

Reform results in Turkey have been shown to be relatively successful, including the following:

- Unbundling the sector into its different business activities (transmission, generation, distribution, wholesale trading, and retail supply)
- Restructuring the existing state-owned entities into independent corporate entities, thus diversifying the numbers of sellers and buyers
- Creating an independent energy regulator (EMRA) and implementing a regulatory framework and a licensing regime
- Making progress in privatizing the state-owned distribution and generation businesses

The introduction of a cost-based pricing mechanism is a significant achievement. However, cross-subsidies remain sizable and in some cases serve as a political tool. Government interference in the affairs of independent energy agencies also remains a destabilizing factor.

Annex 11.1 Turkey Case Study Figures

INCOME LEVEL: Upper-middle-income
REGION: Europe and Central Asia
ENERGY NET IMPORTER/EXPORTER: Net importer
SUBSIDIES: Electricity
PHASING OUT SUBSIDIES: Successful

Fiscal Burden of Energy Subsidy in Turkey

Figure 11A.1   Implicit Subsidies of the Power Sector in Turkey, 2000–03

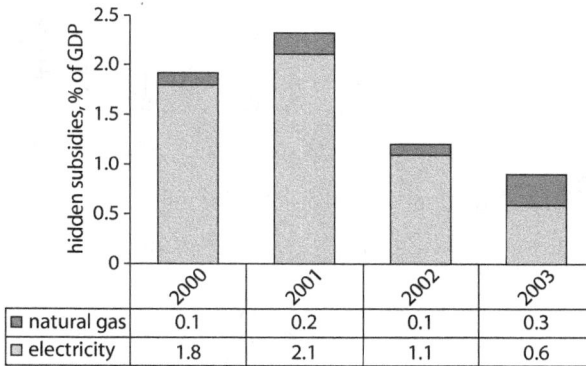

	2000	2001	2002	2003
natural gas	0.1	0.2	0.1	0.3
electricity	1.8	2.1	1.1	0.6

Source: Ebinger 2006.
Note: Implicit subsidies (or hidden costs) are defined as the difference between actual receipts and the revenue that the energy company (for example, a utility involved in the distribution of electricity and natural gas) would receive were it to be in operation with cost-recovery tariffs based on efficient operation with normal losses and with full bill collection.

Fuel Prices and Road Sector Consumption in Turkey

Figure 11A.2   Domestic Retail Fuel Prices in Turkey, 2002–10

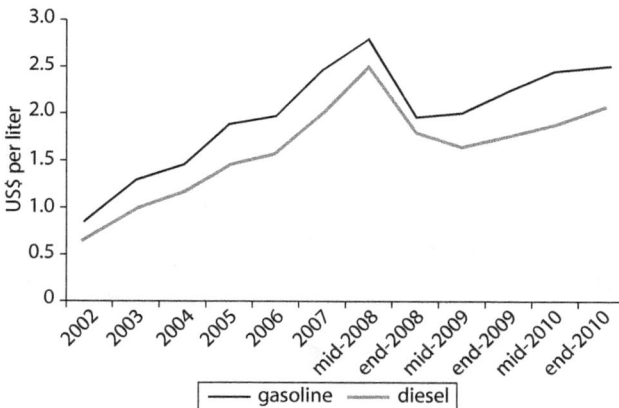

Source: EMRA.

Figure 11A.3   Road Sector Diesel Consumption in Turkey, 1998–2008

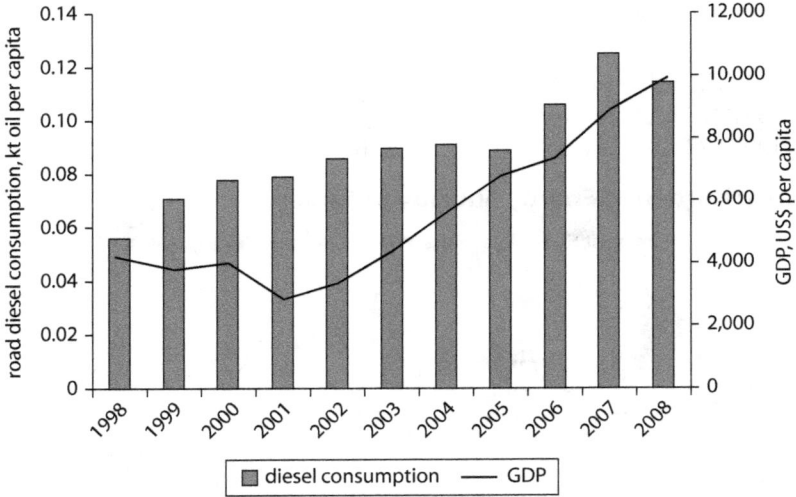

Source: World Bank, World Development Indicators.

Figure 11A.4   Road Sector Gasoline Consumption in Turkey, 1998–2008

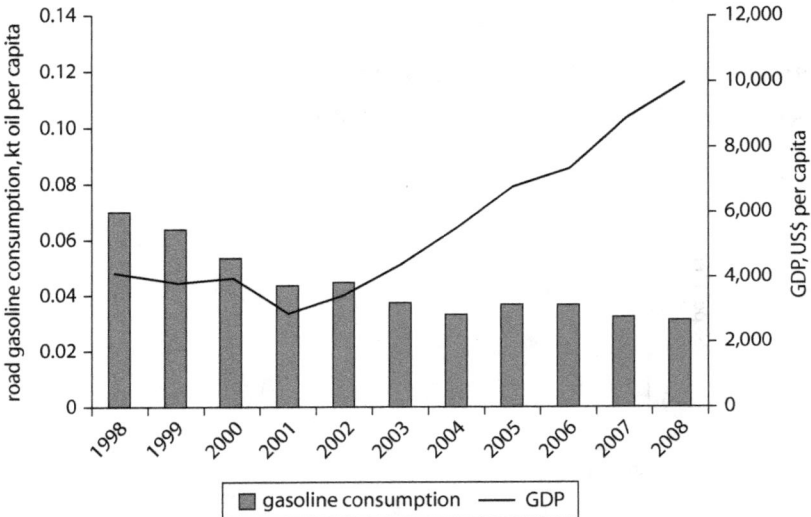

Source: World Bank, World Development Indicators.

Electricity Price and Power Consumption in Turkey

Figure 11A.5   Electricity Price in Turkey, 1998–2010

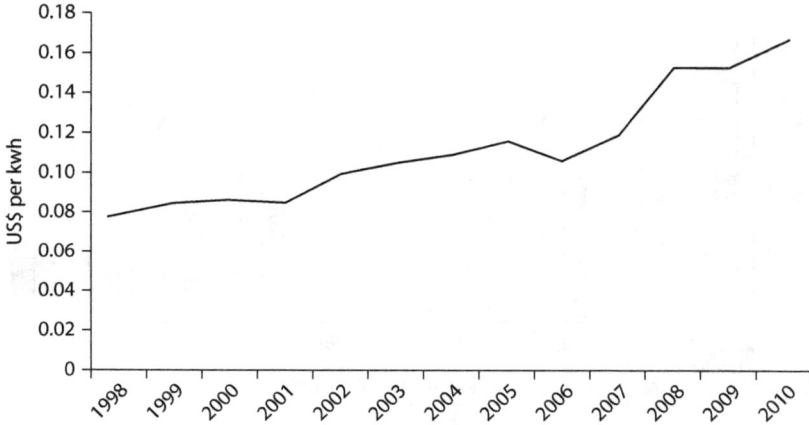

Source: EMRA.
Note: kWh = kilowatt-hour.

Figure 11A.6   Power Consumption Per Capita in Turkey, 1998–2008

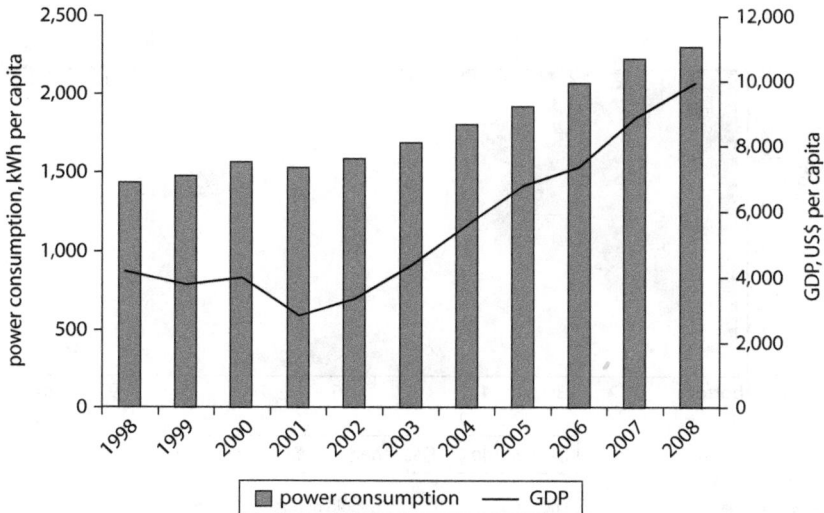

Source: World Bank, World Development Indicators.
Note: kWh = kilowatt-hour.

Poverty Impact Evidence from Household Surveys in Turkey

Figure 11A.7   Energy Expenditure in Turkey, by Income Decile, 2005

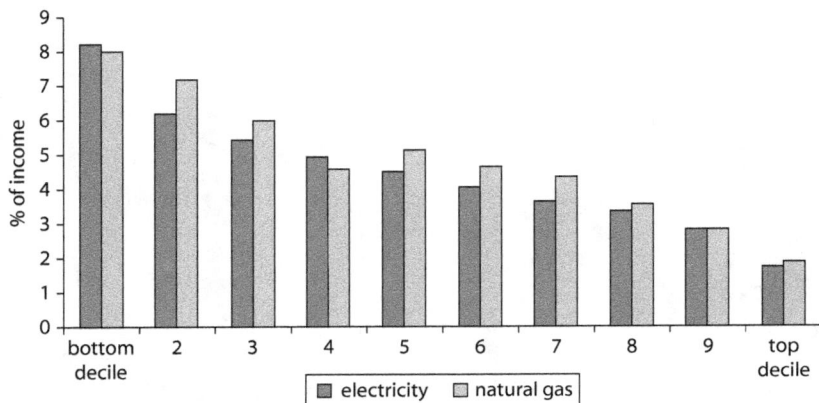

Source: 2005 household budget expenditure.

Figure 11A.8   Welfare Effect under Three Scenarios of Subsidy Reform in Turkey, by Income Decile, 2005

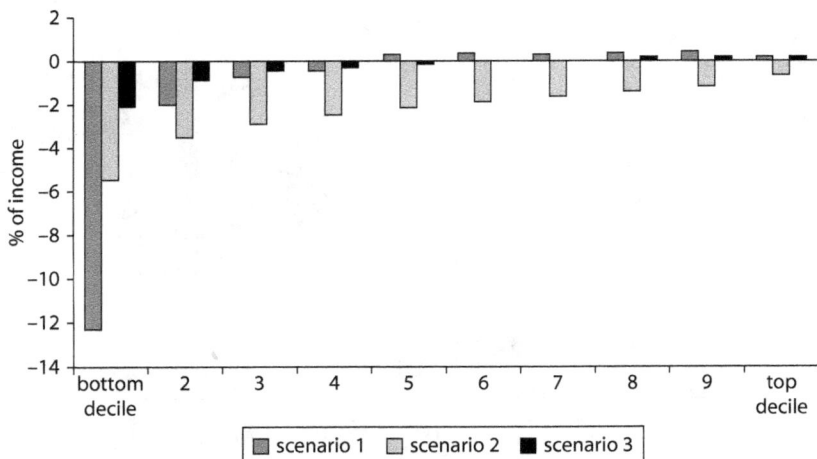

Source: Bagdadioglu, Basaran, and Price 2008, from 2005 household budget expenditure survey.
Note: "Scenario 1" simulates tariffs that charge regions proportionally more for losses they incur. "Scenario 2" simulates a tariff increase for residential users. "Scenario 3" simulates a revenue-neutral standing increase of 10 percent to the average bill.

Figure 11A.9 Electricity Expenditure in Turkey, by Income Quintile, 2003–08

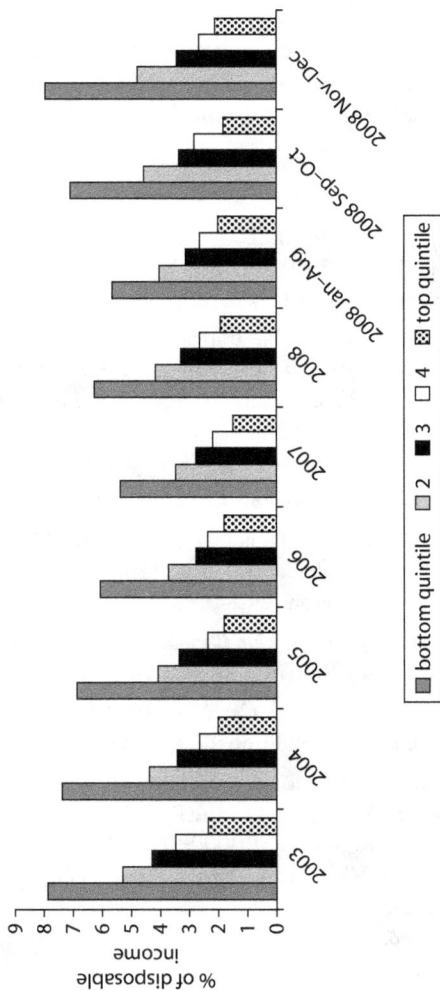

Legend: ☐ bottom quintile ☐ 2 ■ 3 ☐ 4 ☒ top quintile

y-axis: % of disposable income (0–9)

x-axis categories: 2003, 2004, 2005, 2006, 2007, 2008, 2008 Jan–Aug, 2008 Sep–Oct, 2008 Nov–Dec

Source: 2003–08 household budget surveys (World Bank 2010).

Notes

1. Law No. 5784.
2. The CCT program raised secondary school enrollment for girls by 10.7 percent. In secondary schools, education transfers from the CCT program also raised girls' attendance rates by 5.4 percentage points. The CCT program raised primary school attendance for girls by 1.3 percentage points, but there was no positive impact on primary school enrollment rates. There was also no evidence that the CCT program affected the rate of progression from primary school to secondary school (World Bank 2008).

References

Alper, C. Emre, and Orhan Torul. 2009. "Asymmetric Adjustment of Retail Gasoline Prices in Turkey to World Crude Oil Price Changes: The Role of Taxes." *Economics Bulletin* 29 (2): 775–87.

Bagdadioglu, Necmiddin, Alparslan Basaran, and Catherine Waddams Price. 2008. "Potential Impact of Electricity Reforms on Turkish Households." Working Paper 07-8, Centre for Competition Policy, Economic and Social Research Council, University of East Anglia, Norwich, U.K.

Cetin, Tamer, and Fuat Oguz. 2007. "The Politics of Regulation in the Turkish Electricity Market." *Energy Policy* 35 (3): 1761–70.

Ebinger, Jane. 2006. "Measuring Financial Performance in Infrastructure: An Application to Europe and Central Asia." Policy Research Working Paper 3992, World Bank, Washington, DC.

Fiszbein, A., and N. Schady. 2009. *Conditional Cash Transfers: Reducing Present and Future Poverty*. Policy Research Report. Washington, DC: World Bank.

Gokalp, Yadigar. 2007. "Conditional Cash Transfers in Turkey." Presentation at World Bank Workshop on Conditional Cash Transfers, Rabat, Morocco.

World Bank. 2008. "Implementation Completion and Results Report on a Loan in the Amount of US$500 Million to the Republic of Turkey for a Social Risk Mitigation Project." Report ICR0000306, Human Development Unit, Country Department VI, Europe and Central Asia Region, World Bank, Washington, DC. June 26.

———. 2010. "Second Programmatic Environmental Sustainability and Energy Sector Development Policy Loan." IBRD program document, Sustainable Development Unit, Turkey Country Unit, Europe and Central Asia Region, World Bank, Washington, DC.

———. n.d. "CCT Program Profile: Turkey." http://web.worldbank.org/WBSITE/EXTERNAL/TOPICS/EXTSOCIALPROTECTION/EXTSAFETYNETSANDTRANSFERS/0,,contentMDK:20863920~pagePK:148956~piPK:216618~theSitePK:282761~isCURL:Y,00.html.

PART 3

Group C Countries:
Net Energy Exporter and Low Income

Macroeconomic and Social Challenges

- The majority of countries in Group C are characterized by an increasing level of income, as displayed by a high rate of growth in gross domestic product (GDP) per capita as well as a reduction in the level of income inequality, with the notable exception of the Republic of Yemen. Azerbaijan displays the highest GDP per capita and one of the lowest Gini indexes in Group C. Indonesia and the Islamic Republic of Iran follow a similar pattern of a high rate of growth accompanied by significant reduction in inequality (see figures P3.1 and P3.2).
- All countries have been characterized by a declining budget deficit and public debt. However, Azerbaijan and Nigeria are the only countries in Group C that managed to have a budget surplus. Indonesia made significant headway in getting its budget deficit under control and reduced its public debt by a third (see figures P3.3. and P3.4).

Fossil Fuel Dependence

- All countries in Group C have either increased or kept constant the percentage of electricity generated from fossil fuels, but they significantly decreased their net exports of energy over time, with the notable exceptions of Azerbaijan and Indonesia (see figures P3.5 and P3.6).

- All countries except for Nigeria rely almost entirely on electricity generated from fossil fuels as the primary energy source. Only Azerbaijan and Indonesia increased their net energy exports over time. Azerbaijan is characterized by the highest increase in net energy exports—from about 40 percent to more than 340 percent of its own use. Nigeria is characterized by the highest percentage increase of fossil fuel use for electricity production, bringing it up from 62 percent in 1998 to 72 percent in 2008. At the same time, Nigeria recorded a small decrease in net energy exports—from more than 120 percent to about 100 percent of its own use.

Income and Inequality Trends for Group C

Figure P3.1   GDP Per Capita, Group C Countries, 1998–2008

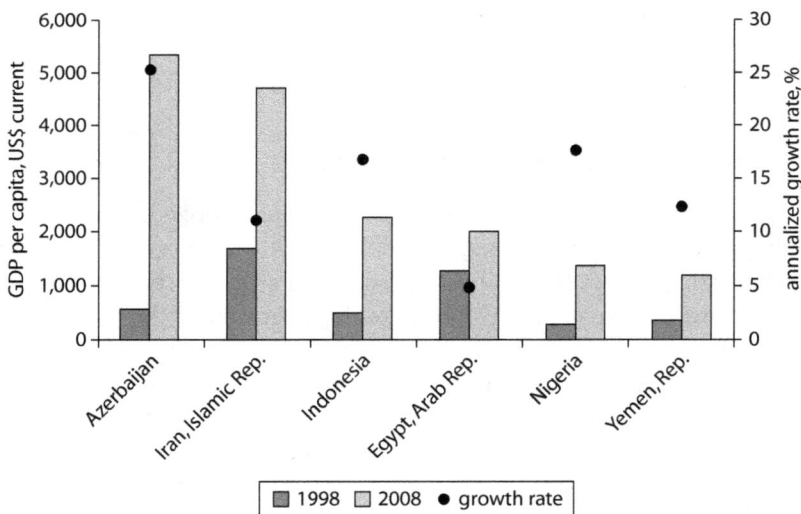

Source: World Bank, World Development Indicators.

Figure P3.2   Gini Index, Group C Countries, 1998–2008

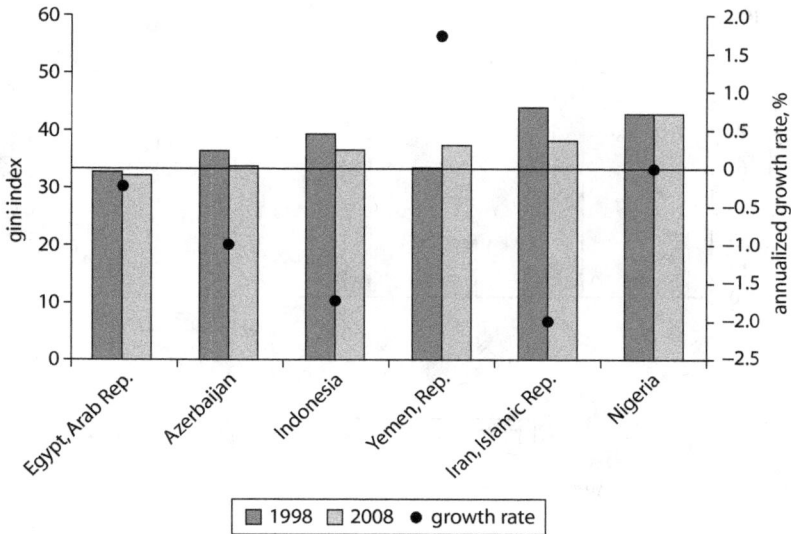

Source: World Bank, World Development Indicators.
Note: The Gini index measures the extent to which the distribution of income (or consumption expenditure) among individuals or households within an economy deviates from a perfectly equal distribution. A Gini index of 0 represents perfect equality, while an index of 100 implies perfect inequality.

Fiscal Indicators for Group C

Figure P3.3   General Government Net Lending or Borrowing, Group C Countries, 1998–2008

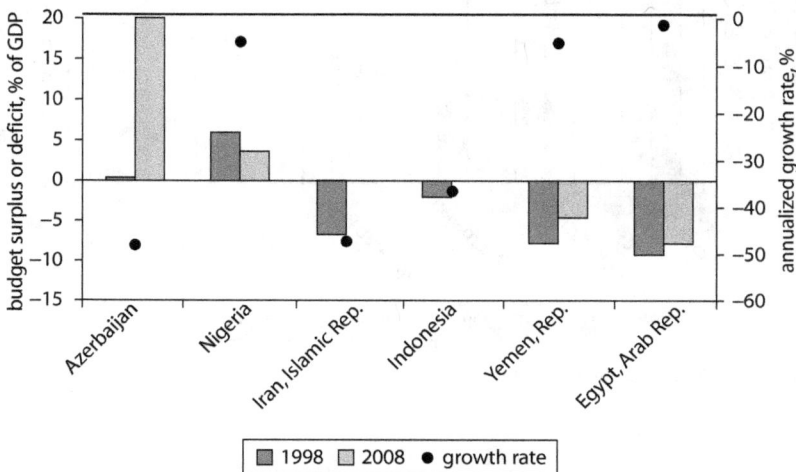

Source: IMF reports, various years.

Figure P3.4   General Government Gross Debt, Group C Countries, 1998–2008

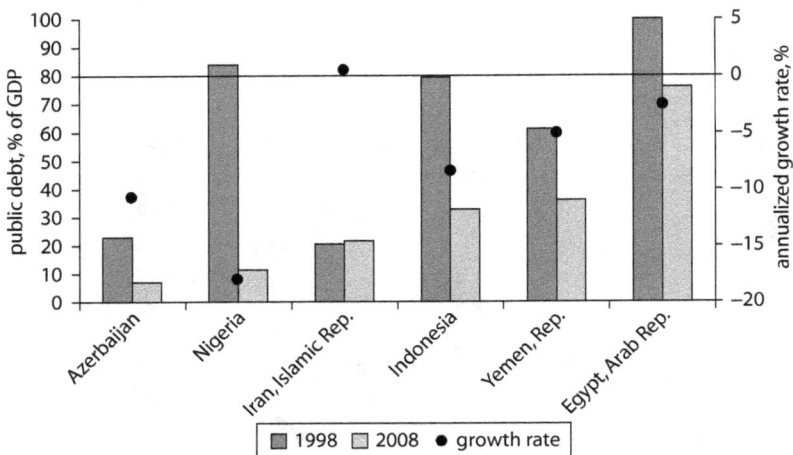

Source: IMF reports, various years.

Fossil Fuel Dependence for Group C

Figure P3.5   Electricity Production from Fossil Fuels, Group C Countries, 1998–2008

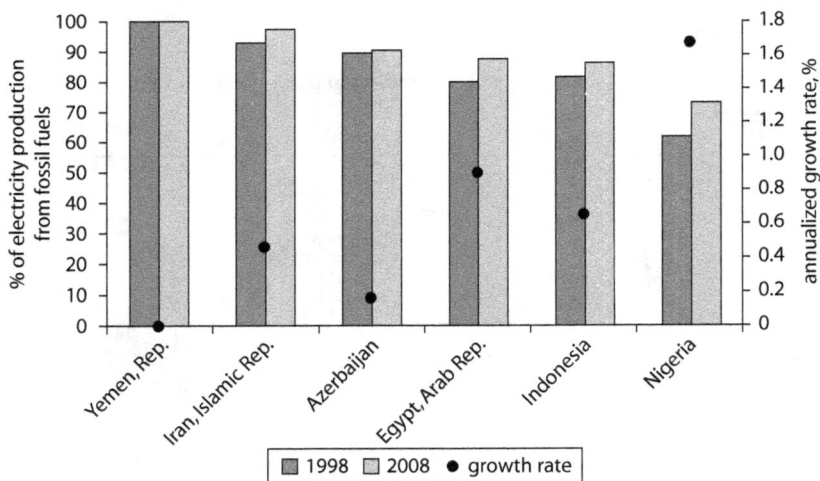

Source: World Bank, World Development Indicators.

Figure P3.6   Energy Net Imports, Group C Countries, 1998–2008

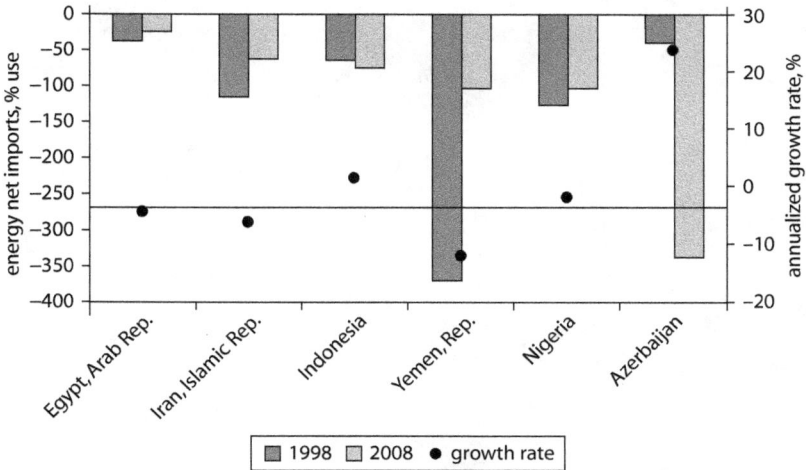

Source: World Bank, World Development Indicators.

CHAPTER 12

Azerbaijan

Incentives to Energy Subsidy Reforms

Fossil Fuels

Extractive industries (oil and gas) account for the majority of industrial output and export revenues of Azerbaijan. Their importance to the economy has increased over time to around 80 percent of industrial output in 2009, up from 50 percent in 2004. Earnings from hydrocarbon sales constituted almost 93 percent of total export revenue in 2009. Azerbaijan has only a small share of global hydrocarbon reserves, but it has ramped up oil production since 2005 following the development of the offshore Azeri-Chirag-Guneshli fields, accounting for a large share of non-OPEC (Organization of the Petroleum Exporting Countries) supply growth. Almost all of this oil is exported via the Baku-Tbilisi-Ceyhan pipeline system. Although the oil boom has generated opportunities for economic and social development, it is highly temporary in nature (IMF 2010a). Oil production is expected to peak in 2014, and oil reserves are expected to be exhausted in 20–25 years unless new discoveries are made.

Azerbaijan established the Azeri State Oil Fund (SOFAZ) in 1999 and in 2003 joined the Extractive Industries Transparency Initiative. As of July 2010, the fund's assets reached US$18 billion. Despite the large drop in oil prices, the budgeted transfer of resources from SOFAZ was fully implemented in 2009 (see figure 12.1).

The State Oil Company of the Azerbaijan Republic (SOCAR) is involved in all oil and gas projects in Azerbaijan. SOCAR is also responsible for operating the refineries, running the pipeline network (except the Baku-Tbilisi-Ceyhan pipeline), and managing oil and gas exports. Due to the lower-than-expected oil price in 2009, the government provided a capital injection and government-guaranteed loans to SOCAR and to the state-owned aluminum company, amounting to 3.2 percent of gross domestic product (GDP) (IMF 2010a).

The 2007 oil price increase (further discussed in the "Reform Efforts" section below) helped in reducing explicit subsidies in the 2006 budget and eventually removing them from the budget in 2007 (see figure 12A.1). Implicit subsidies also declined while remaining sizable given the increasing gap between domestic energy prices and world prices since January 2007. Costs for adjustments remain large at about 15 percent of nonoil GDP as of 2008, according to International Monetary Fund estimates (figure 12A.2).

Electricity

Historically, a number of subsidies were embedded in electricity tariffs in Azerbaijan as a result of both nonpayment problems in the sector and below-cost pricing (see figure 12A.3). In 2003, it is estimated that the shortfall in payments accounted for about 3.2 percent of GDP. The underrecovery of the true economic value of the fuels supplied to Azerenergy was equal to 2.5 percent. These are the two

Figure 12.1   Trends in the Azeri State Oil Fund (SOFAZ), Azerbaijan, 2005–09

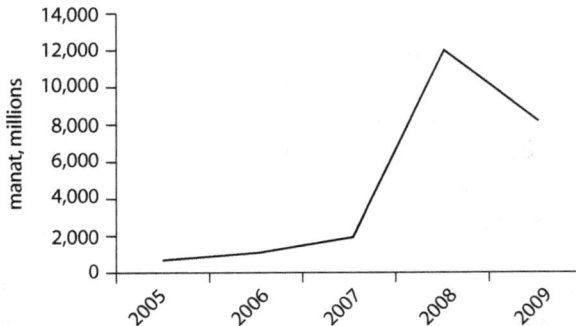

Source: IMF 2010a.

largest components of the hidden costs shown in figure 12A.3. SOCAR provided electric and gas utilities (Azerenergy, Azerigas, and Azerchemia) with fuel sufficient for their operations even when the companies could not make their payments in full. Since 2003, the government started to reflect these underpayments as subsidies in the budget.

Reform Efforts

Fossil Fuels

The Tariff Council was established by a presidential decree in 2005 as a separate agency responsible for setting prices and tariffs for a wide range of goods and services, including electricity, gas, and petroleum products.

One of the first actions of the newly-established council was the doubling of prices of diesel, kerosene, and gasoline in January 2006 (see figure 12A.4). The price of diesel doubled and reached the level of the gasoline price. The price of kerosene increased by a factor of 2.3. The increases reflected the policy of the government to reduce the gap between domestic prices and world market prices. The price hikes were expected to increase state budget revenues by about US$85 million and to reduce smuggling of fuels into neighboring Georgia and the Russian Federation, where prices were higher.

The next price increase took place in January 2007. This time, along with the increases in the prices of petroleum products, there was also a dramatic increase in the price of electricity, which went up almost three-fold—the first increase in the electricity tariff since 1997 (see figure 12A.7). The 2007 price increases were expected to bring an additional US$220 million in budget revenues.

Road diesel consumption declined substantially in 2006 and 2007, then dropped to 2004 levels in 2008, whereas gasoline consumption by the road sector continued to increase, fueled by GDP growth (see figures 12A.5 and 12A.6).

Electricity

In January 2007, as previously mentioned, Azerbaijan passed through substantial price increases (coming from the increased cost of generation) to its electricity consumers. The government tripled electricity prices for consumers in an effort to phase out subsidies to the sector (see figure 12A.7). Natural gas had been increased by 650 percent in the two years prior to electricity tariff reforms (Mehta, Rao, and Terway 2007).

Due to strong oil exports in 2007, the Azerbaijani economy still grew by a staggering 23 percent, yet the power consumption decreased by 5 percent. In 2008, power consumption decreased by another 4 percent (see figure 12A.8). The drop in electricity demand was due to the price increases in January 2007 as well as the abrupt cessation of Russian gas imports at the end of 2006.

Price reforms were accompanied by comprehensive electricity sector rehabilitation. After years of underinvestment, the authorities have now started to use export revenue to upgrade the electricity generating and distribution networks. The sector is already showing improvements. A comprehensive electricity metering program for all energy users was implemented in 2007, aimed at enhancing collection rates and improving energy efficiency (IEA 2010).

Poverty Alleviation Measures

Electricity
Azerbaijan is one of the few countries that did not adopt increasing block tariffs. There have been in place some discounts to special customer categories, including internally displaced persons (IDPs) in Azerbaijan. Although there are conflicting data on the incidence of poverty within this group of the population, mitigating the potential blow from tariff increases is a government priority. Currently IDPs receive an allowance, paid by the State Refugee Committee to Barmek and Bayva, of 150 kilowatt-hours per person per month (Lampietti, Banerjee, and Branczik 2007). Collection rates are significantly higher for the top income quintile (see figure 12A.9).

Evidence from Household Surveys
The availability of two household surveys in 2001 and 2008 allows us to compare the impact of price reforms on household expenditure. After the price hikes of 2006 and 2007, the poorest households spent a considerably higher percentage of their incomes on utilities, which indicates that consumption did not sharply decline, whereas the top quintile kept spending the same percentage of income, which may also reflect some reduction in consumption (see figure 12A.10).

Social Safety Nets
Azerbaijan has a good track record of reforming its social protection system and policies. Ongoing reforms aim to put in place an improved social

risk management mechanism through strengthened social protection institutions, improved service delivery, and efficient targeting. So far, the focus has been on streamlining cash transfer programs, improving their fiscal sustainability, and strengthening their poverty focus as well as on improving the transparency and efficiency of the social insurance and targeted social assistance (TSA) administrations (World Bank 2008).

The TSA program is aimed at poverty alleviation among poor households—covering the difference between per capita household income and the subsistence minimum—and was launched in mid-2006. The targeting method is based on income testing, which is obtained by visiting each applicant household in order to determine its eligibility (World Bank 2008). Preliminary assessments of the TSA program suggest that the targeted level of the program was higher than that of other programs. Of the TSA beneficiaries, 86 percent were from the poorest two income quintiles. Estimations indicate that, in the absence of social assistance, the poverty incidence could have increased from 10.8 percent up to 21 percent and the poverty gap from 2.4 percent up to 7.2 percent (World Bank 2008).

Key Lessons Learned

Azerbaijan has made significant progress in reducing explicit subsidies. Implicit subsidies still remain large, but these are also expected to drop as electricity sector operations improve. Payment collections have also improved because of better service and implementation of a metering system.

Azerbaijan represents a successful case of energy sector reform consisting of reduction of explicit subsidies carried out in tandem with electricity sector rehabilitation and improvement in the quality of services to customers. In the case in which tariffs are increased but service quality is not improved, it will create incentives for customers either to reduce consumption or to increase theft or nonpayment—creating a vicious circle that will, in turn, seriously limit the ability to generate revenues for utilities to invest in the sector and improve the quality of service. Efforts in streamlining cash transfer programs and strengthening targeting methods are also praiseworthy.

Annex 12.1 Azerbaijan Case Study Figures

INCOME LEVEL: Lower-middle income
REGION: Europe and Central Asia
ENERGY NET IMPORTER/EXPORTER: Net exporter
SUBSIDIES: Electricity, gas
PHASING OUT SUBSIDIES: Ongoing

Fiscal Burden of Energy Subsidy in Azerbaijan

Figure 12A.1   Explicit Budgetary Oil Subsidies in Azerbaijan, 2003–10

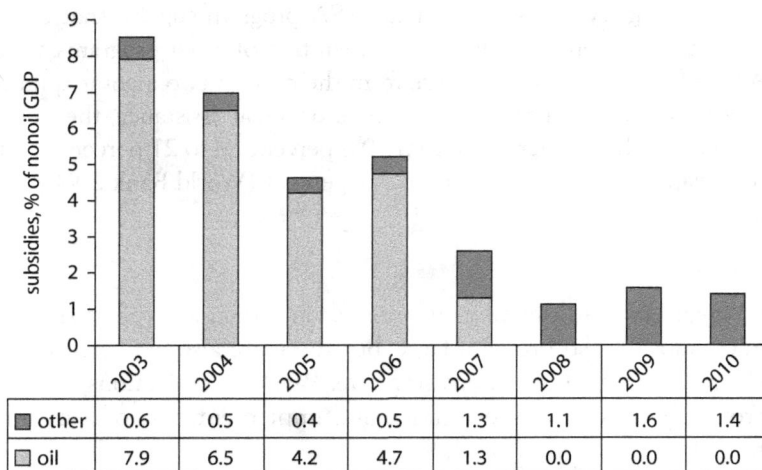

	2003	2004	2005	2006	2007	2008	2009	2010
■ other	0.6	0.5	0.4	0.5	1.3	1.1	1.6	1.4
□ oil	7.9	6.5	4.2	4.7	1.3	0.0	0.0	0.0

Source: IMF staff reports, various years.

Figure 12A.2   Implicit Oil Subsidies in Azerbaijan, 2003–08

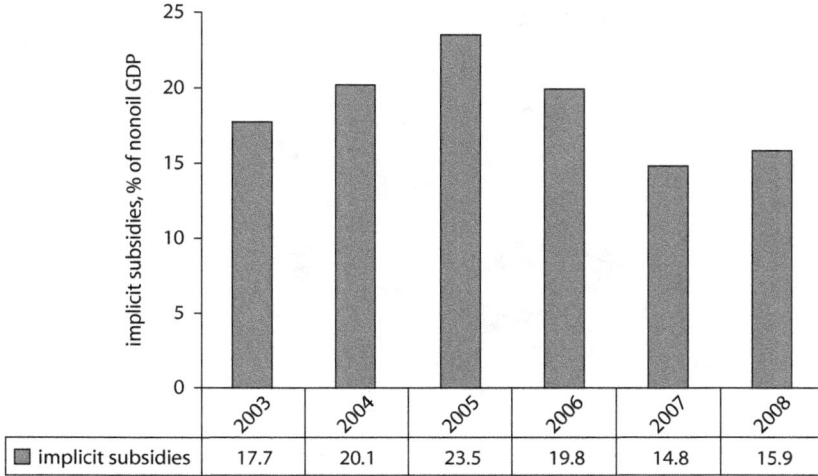

	2003	2004	2005	2006	2007	2008
implicit subsidies	17.7	20.1	23.5	19.8	14.8	15.9

y-axis: implicit subsidies, % of nonoil GDP

Source: IMF staff reports, various years.

Figure 12A.3   Implicit Subsidies in the Power Sector in Azerbaijan, 2000–03

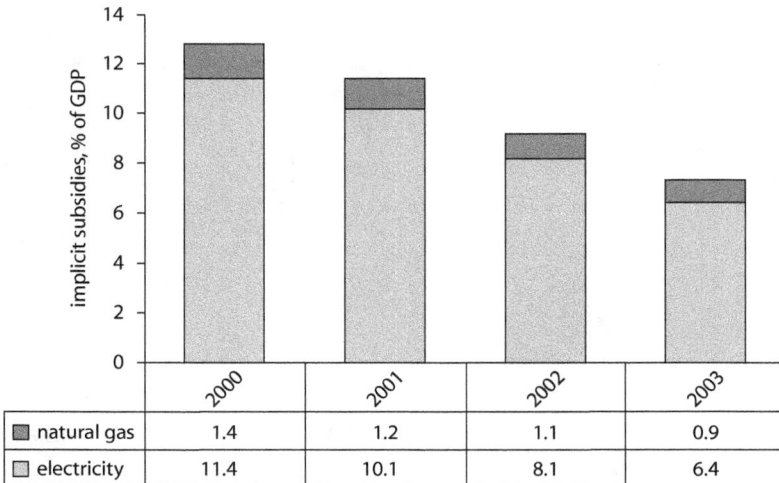

	2000	2001	2002	2003
natural gas	1.4	1.2	1.1	0.9
electricity	11.4	10.1	8.1	6.4

y-axis: implicit subsidies, % of GDP

Source: Ebinger 2006.
Note: Implicit subsidies (or hidden costs) are defined as the difference between actual receipts and the revenue that the energy company (for example, a utility involved in the distribution of electricity and natural gas) would receive were it to be in operation with cost-recovery tariffs based on efficient operation with normal losses and with full bill collection.

Fuel Prices and Road Sector Consumption in Azerbaijan

Figure 12A.4   Domestic Retail Fuel Prices in Azerbaijan, 2002–10

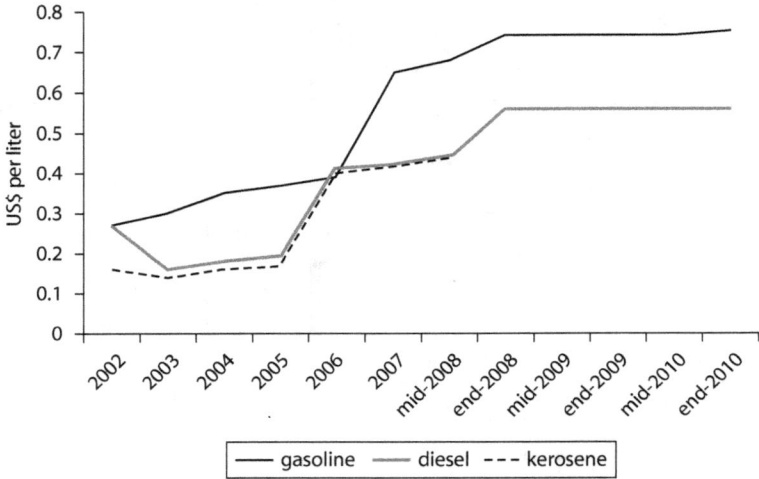

Source: Elaboration of data from GIZ n.d.; IMF 2010b; and additional data from individual country information.

Figure 12A.5   Road Sector Diesel Consumption in Azerbaijan, 1998–2008

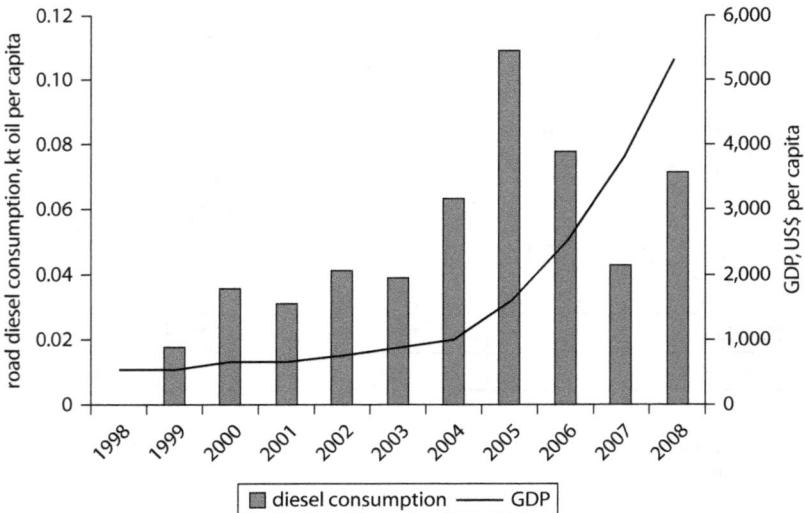

Source: World Bank, World Development Indicators.

Figure 12A.6   Road Sector Gasoline Consumption in Azerbaijan, 1998–2008

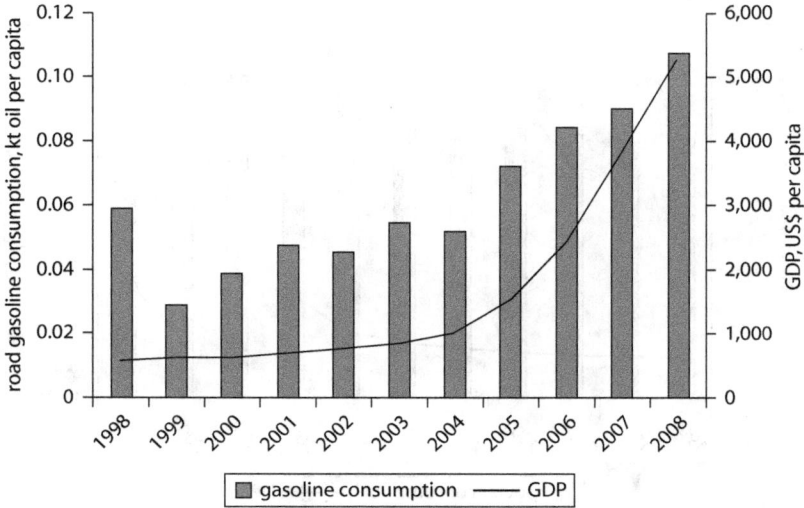

Source: World Bank, World Development Indicators.

Electricity Price and Power Consumption in Azerbaijan

Figure 12A.7   Electricity Price in Azerbaijan, 2000–10

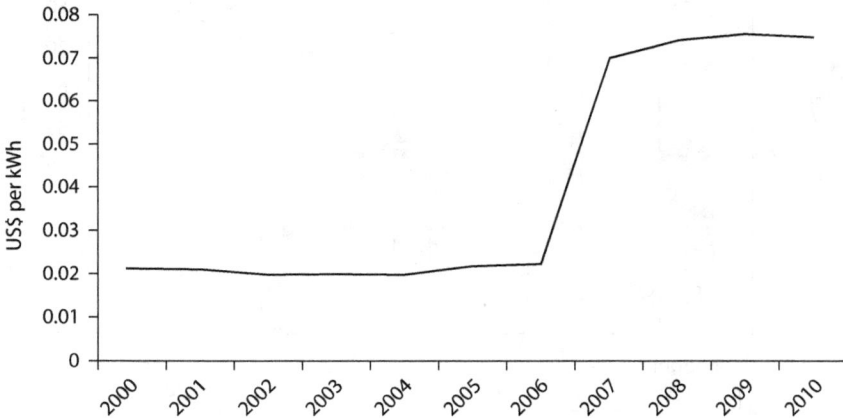

Source: Azerbaijan Tariff Council.
Note: kWh = kilowatt-hour.

Figure 12A.8   Power Consumption Per Capita in Azerbaijan, 1998–2008

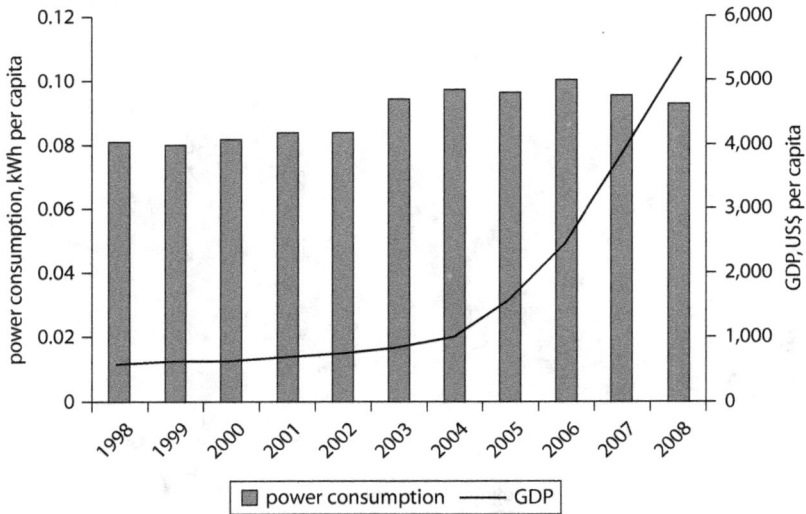

Source: World Bank, World Development Indicators.
Note: kWh = kilowatt-hour.

Poverty Impact Evidence from Household Surveys in Azerbaijan

Figure 12A.9   Electricity Expenditure and Collection Rate in Azerbaijan, by Income Quintile, 2001

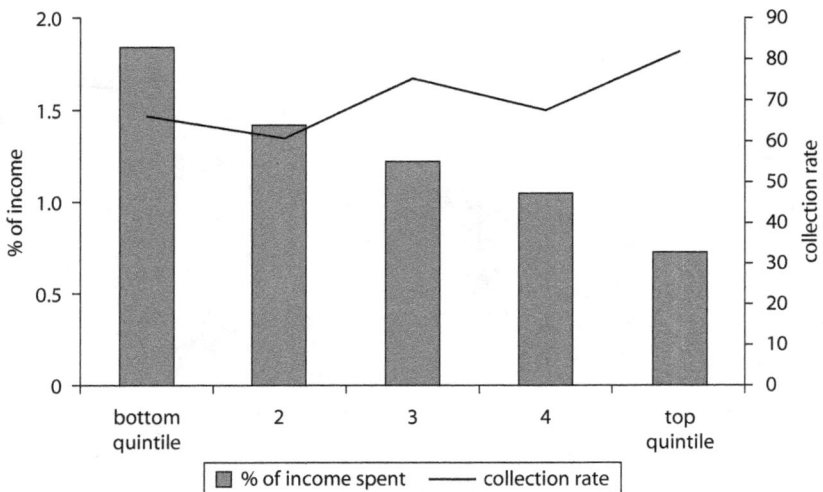

Source: World Bank 2004, based on 2001 household budget survey.

Figure 12A.10   Utility Expenditure in Azerbaijan, by Income Quintile, 2001–08

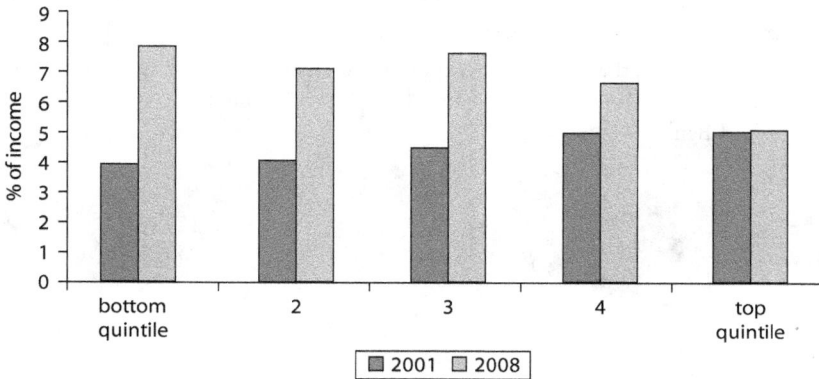

Source: World Bank 2010, based on 2001 and 2008 household budget surveys.
Note: Utilities include electricity, heating, water, sewerage, and communal services.

References

Ebinger, Jane. 2006. "Measuring Financial Performance in Infrastructure: An Application to Europe and Central Asia." Policy Research Working Paper 3992, World Bank, Washington, DC.

GIZ (German Agency for International Cooperation). n.d. International Fuel Prices database. GIZ (formerly GTZ), Bonn. http://www.gtz.de/en/themen/29957.htm.

IEA (International Energy Agency). 2010. *World Energy Outlook 2010.* Paris: Organisation for Economic Co-operation and Development/IEA.

IMF (International Monetary Fund). 2010a. "Azerbaijan: Article IV Consultation." IMF, Washington, DC.

———. 2010b. Retail domestic fuel prices data sheet, IMF, Washington, DC. http://www.imf.org/external/pubs/ft/spn/2010/data/spn1005.csv.

Lampietti, Julian A., Sudeshna Ghosh Banerjee, and Amelia Branczik. 2007. *People and Power: Electricity Sector Reforms and the Poor in Europe and Central Asia.* Washington, DC: World Bank.

Mehta, Aashish, H. Satish Rao, and Anil Terway. 2007. "Power Sector Reform in Central Asia: Observations on the Diverse Experiences of Some Formerly Soviet Republics and Mongolia." *Journal of Cleaner Production* 15 (2): 218–34.

World Bank. 2004. "Azerbaijan Raising Rates: Short-Term Implications of Residential Electricity Tariff Rebalancing." Report 30749-AZ, Europe and Central Asia Region, Environmentally and Socially Sustainable Development, World Bank, Washington, DC.

———. 2008. "Social Protection in Azerbaijan." Project Appraisal Document, Human Development Sector Unit, South Caucasus Country Department, Europe and Central Asia Region, World Bank, Washington, DC.

———. 2010. "Azerbaijan Living Conditions Assessment Report." Report 52801-AZ, Human Development Sector Unit, Europe and Central Asia Region, World Bank, Washington, DC.

CHAPTER 13

Arab Republic of Egypt

Incentives to Energy Subsidy Reforms

The Arab Republic of Egypt has significant energy resources but is no longer a net energy exporter. The extraction industries, including oil refining and natural gas, accounted for 13.3 percent of nominal gross domestic product (GDP) in fiscal year 2008/09, with the electricity sector representing another 1.4 percent. Oil production has gradually fallen as reserves have dwindled, although new discoveries have boosted total reserves in recent years. Egypt also has substantial natural gas reserves. The government is promoting the consumption of natural gas over oil domestically and is now the third-biggest gas exporter in the Middle East (after Qatar and Algeria, in that order).

Egypt's government has set the goal of having 20 percent of its installed generation capacity in the form of renewable energy by 2020. It is already a regional leader in wind-power generation, with an installed 500 megawatts (MW) of wind-energy capacity that is performing well. A further 720 MW of projects to expand this program are under way. The World Bank is also implementing an Integrated Solar Combined Cycle Power Project and concentrated solar power technology to increase the share of solar-based electricity in Egyptian energy generation (CTF 2009).

Energy subsidies represent a substantial fiscal drain on the Egyptian economy. Since the 2005/06 fiscal year—the first time in which subsidies for petroleum (oil and gas) fuels started to be accounted for in the budget—subsidies in Egypt increased from LE 40 billion (equivalent to about US$7.2 billion) to LE 66 billion (equivalent to US$11.8 billion) in the 2009/10 fiscal year and were expected to hike even further to LE 82 billion (equivalent to about US$14 billion) in 2011. Petroleum subsidies represent an annual average above 6 percent of GDP and account for the most significant source of budgetary subsidies (see figure 13A.1). The budgetary subsidies, as high as they are, significantly underestimate the real economic cost of subsidies because they record only financial subsidies and do not account for the cost of economic distortions and inefficiencies caused by the significant underpricing of energy supply. A recent study commissioned by the World Bank, undertaken for the government of Egypt and financed by the Energy Sector Management Assistance Program, estimates that if subsidies were calculated on the basis of full economic costs, the subsidy figure for 2009 would be much higher. For example, in 2009, the energy subsidies would have been twice as high: US$23.7 billion or 14.5 percent of GDP (World Bank 2009).

Natural gas is still used primarily for electricity generation (64 percent) while only 3 percent is directly consumed by households. In 2006-07, Egypt supplied about 78 percent of the total domestic demand, with the remainder purchased locally from foreign partners (Abouleinein, El-Laithy, and Kheir-El-Din 2009). Egypt has also been struggling to supply liquefied petroleum gas (LPG) to its citizens, which is one if its most heavily subsidized fuels. Households consume 41.1 percent of LPG (Abouleinein, El-Laithy, and Kheir-El-Din 2009).

The subsidy phaseout would be timely since Egypt is no longer the plentiful oil producer it used to be in the past. In 2007–08, natural gas earnings exceeded those from oil output as Egypt increased its gas exports. Demand for energy has been growing at an average of 8.1 percent annually between 2007 and 2009—which, at this rate, implies adding about 1,500–2,000 MW per year over the next several years to the existing installed capacity of 22,000 MW. There are even worries that the existing natural gas reserves may not be enough to meet future demands and that Egypt may soon become an energy importer. The state gas company decided not to sign new export contracts in 2010 in order to accommodate domestic consumption. In light of this, Egypt will need to consider phasing out subsidies definitively to curb uneconomic domestic energy demand growth.

Reform Efforts

Fossil Fuels

The 2004 cabinet increased prices of gasoline and of diesel. Further increases in fuel prices took place as part of efforts to lower the subsidy bill since January 2008 (see figure 13A.2). The government also decided to introduce a quota-based system for the supply of LPG from September 2010. LPG is heavily subsidized and is used by the majority of Egyptians for domestic heating and cooking. Egypt imports about half of its annual requirement (about 4.2m tons last year). LPG is distributed at an official price of LE 3 (US$0.55) per canister, but the actual cost is about LE 50.

Road fuel consumption has largely been unaffected by such price increases (with the notable exception of a temporary decline in diesel consumption in 2005), and consumption generally continued to grow, fueled by increases in GDP (see figures 13A.3 and 13A.4).

Electricity

As a result of a process of unbundling in 2000, Egypt is now served by nine electricity distribution companies as well as six generation companies and one transmission company, all of which are affiliated and controlled by their parent company: an Egyptian joint stock (holding) company under the name Egyptian Electricity Holding Company (EEHC). EEHC is also responsible for the planning, development, and operation of the government-owned electric utilities. At present, the Egyptian Electric Transmission Company (EETC) operates under a single-buyer model, purchasing bulk power from all generation entities. EETC in turn sells bulk power to distribution companies and to high-voltage (HV) and extra-high-voltage (EHV) customers. EEHC sets and controls the purchase and selling prices of electricity among the government-owned utilities.

All of these reforms took place without an independent regulator. Despite the issuance of Presidential Decree No. 326 in 1997 to establish the Electric Utility and Consumer Protection Regulatory Agency (ERA), ERA did not start operations until early 2002. ERA's powers fall short of a truly independent regulatory agency, however, because it does not have authority over tariff setting, and its rulings are under government influence because its board is chaired by the Minister of Electricity and Energy.

In 2004, Egypt increased its electricity prices for the first time since 1992 (see figure 13A.5). Due to a further spike in demand and a rise in international fuel prices, the government began another reduction of subsidies in the 2007/08 fiscal year, raising the price for 37 energy-intensive domestic companies, including producers of iron and steel, aluminum, cement, and fertilizers but excluding the petrochemicals.

In 2004, the government started a program of increasing natural gas prices for industry, starting with the energy-intensive sectors. Companies classified as energy-intensive pay US$3 per million British thermal units (mmBtu) of natural gas. In June 2008, the government announced that the price of natural gas for energy-intensive industrial users—which account for about 60 percent of industrial consumption of energy—would be increased from US$1.25 per mmBtu to US$3 per mmBtu with immediate effect. Likewise, the electricity price for energy-intensive industrial users was increased in one step to US$0.063 per kilowatt-hour (kWh), US$0.046 per kWh, and US$0.038 per kWh for medium, EHV, and ultra-high-voltage customers, respectively.

Regarding average retail tariffs, the adjustments started in 2004, from an average of US$0.022 per kWh to US$0.024 per kWh in that year, reaching the average of US$0.035 per kWh by 2008. These low electricity prices are only possible due to subsidized natural gas and fuels used for electricity generation. The price of natural gas (and consequently of electricity generated mainly by natural gas) for non-energy-intensive industries was increased from US$1.70 per mmBtu to US$2 per mmBtu in July 2010. This increase was suspended during the global economic downturn, and a further increase in the price of gas for energy-intensive industries was not expected until after the presidential election in mid-2012. Power consumption continued to increase over time due to buoyant GDP growth (see figure 13A.6).

Poverty Alleviation Measures

Evidence from Household Survey Data

According to the results of the 2005 household survey, 57 percent of energy subsidies are currently benefiting the top two income quintiles of households (World Bank 2009). In the case of urban households, the top two quintiles receive an even more staggering three-quarters of the total energy subsidies. Of the subsidies for gasoline used for cars, 92 percent go to the top income quintile and virtually none to the bottom quintile (see figure 13A.8). Only kerosene and LPG display a progressive trend in rural areas. The overall regressive trend, which is common to other countries, is explained by the fact that benefits are conditional upon the purchase of subsidized goods and increase with expenditures, which in turn also increase with income (see figure 13A.9).

However, subsidy reforms pose formidable challenges because the relative adverse impact of energy subsidy removal is expected to be the greatest for the poor, even though the rich receive the highest share of the subsidy. According to the results of the 2005 household survey, energy subsidies represented over 12 percent of household expenditure for the bottom quintile, but 8.6 percent for the top quintile (World Bank 2009).

To improve the targeting of LPG subsidies, the Ministry of Social Solidarity is implementing LPG coupons for the country's 15 million ration card holders. The beneficiaries are divided into families with three or fewer members who receive one cylinder a month and those with four or more members who are entitled to two cylinders a month at the subsidized price. With the current LPG crisis in Egypt, however, the market price has almost quadrupled as the scarce fuel is being traded in the black market.

Social Safety Nets

Egypt's existing in-kind subsidy programs, which form the basis of the current safety net, are costly, ineffective, create market distortions and inefficiencies, and benefit the rich far more than the poor despite high and increased levels of spending (World Bank 2005). The World Bank recommends strengthening the social safety net along the following lines:

- Expanding the cash assistance program to raise the benefit levels and expand the coverage
- Enriching the safety net with better targeting methods like proxy means testing and greater use of geographic targeting in order to direct a substantial fraction of public resources to the intended beneficiaries while minimizing the leakage to the wealthy
- Introducing conditional cash transfer programs that help the poor enhance their human capital.

Fossil Fuels

Abouleinein, El-Laithy, and Kheir-El-Din (2009) analyzed the impact of the gradual phasing out of fuel subsidies in Egypt over a five-year period, from 2008 to 2013. In the first year of simulation, the price increases implemented in May 2008 include increases in the price of natural gas, heavy oil, diesel, and gasoline by 58.33 percent, 100 percent, 46.7 percent, and 28.2 percent, respectively. Using a computable general equilibrium model, the study showed that the elimination of energy subsidies, without any offsetting policy actions, would reduce average annual GDP growth by 1.4 percentage points over the reference period and depress the welfare levels of households at all levels of the income distribution. Inequality would be reduced, however, reflecting the larger welfare impact on households in the richest quintile of the distribution.

The authors also examined alternative scenarios involving either targeted or untargeted transfers that redistribute up to 50 percent of the energy subsidy savings (see figure 13A.11). Untargeted transfers of this magnitude still result in income losses to households at all levels of the income distribution. However, transfers targeted to the poorest two quintiles of the income distribution increase their welfare relative to what it would have been in the presence of energy subsidies, generating a large improvement in income distribution measures. Economic growth is higher with targeted transfers compared with untargeted transfers because a greater portion of the funds is recycled into higher consumption (although growth is still lower than what it would have been in the absence of the reforms).

Key Lessons Learned

The broader strategy for phasing out energy subsidies is currently discussed. The original plan had been to phase out subsidies for electricity and gasoline completely by 2014, with only LPG continuing to be supported. This timeline was suspended during the global economic downturn, but it needs to be reconsidered by the new incoming government.

The government is gradually increasing energy prices for industry in order to reduce the subsidy burden, but these increases are being phased in slowly and are unlikely to have a major impact on demand.

Increasing further the share of renewables in the energy mix, given Egypt's good potential for both solar and wind energy, will help diversify electricity generation and ease some of the pressure from the dwindling fossil fuel resources in the medium to long run.

Annex 13.1 Egypt Case Study Figures

INCOME LEVEL: Lower-middle income
REGION: Middle East and North Africa
ENERGY NET IMPORTER/EXPORTER: Net exporter
SUBSIDIES: Electricity, oil, LPG, natural gas
PHASING OUT SUBSIDIES: Ongoing

Fiscal Burden of Energy Subsidy in Egypt

Figure 13A.1   Explicit Budgetary Energy Subsidies in Egypt, 2002–10

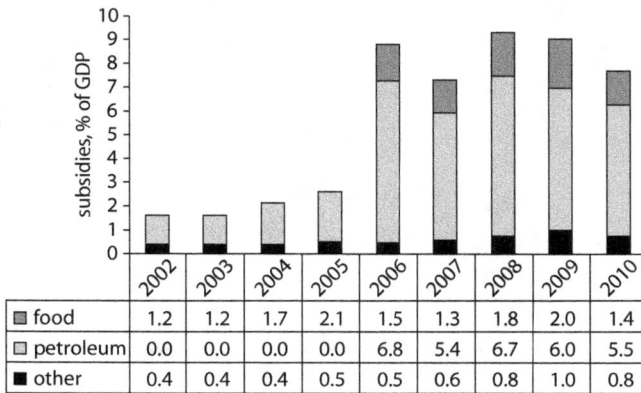

	2002	2003	2004	2005	2006	2007	2008	2009	2010
food	1.2	1.2	1.7	2.1	1.5	1.3	1.8	2.0	1.4
petroleum	0.0	0.0	0.0	0.0	6.8	5.4	6.7	6.0	5.5
other	0.4	0.4	0.4	0.5	0.5	0.6	0.8	1.0	0.8

Sources: Egypt, Ministry of Finance and Central Bank; IMF various years; author's calculations.

Fuel Prices and Road Sector Consumption in Egypt

Figure 13A.2   Domestic Retail Fuel Prices in Egypt, 2002–10

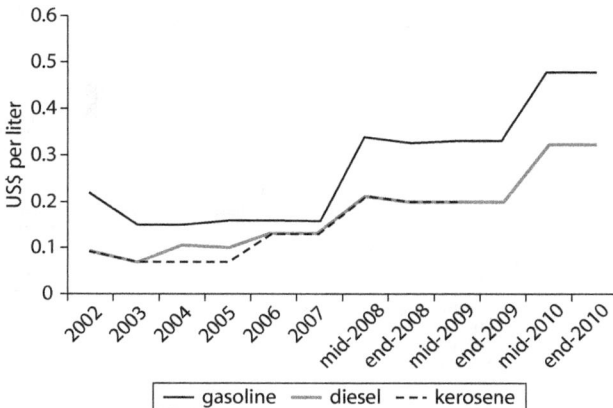

Sources: Elaboration of data from GIZ n.d.; IMF 2010; and additional data from individual country information.

Figure 13A.3   Road Sector Diesel Consumption in Egypt, 1998–2008

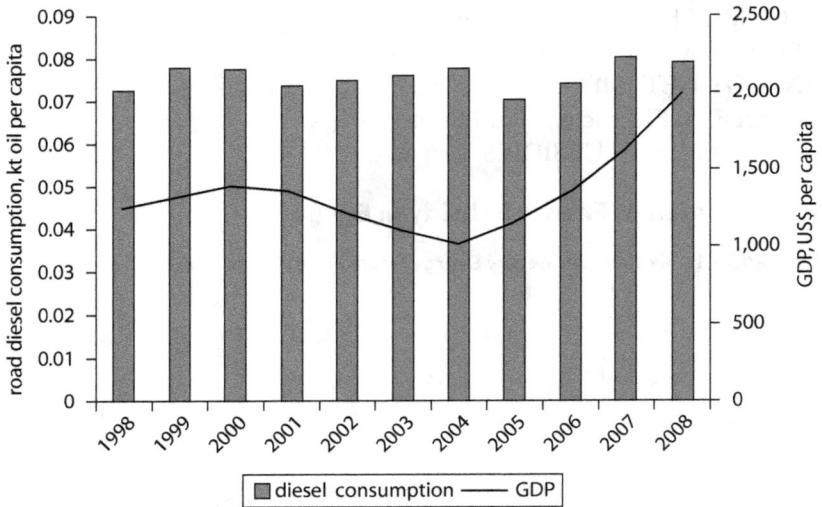

Source: World Bank, World Development Indicators.

Figure 13A.4   Road Sector Gasoline Consumption in Egypt, 1998–2008

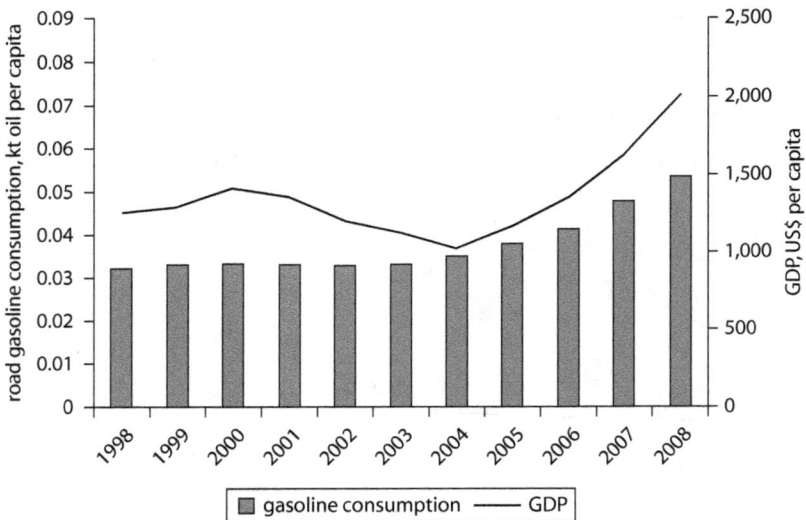

Source: World Bank, World Development Indicators.

Electricity Price and Power Consumption in Egypt

Figure 13A.5   Electricity Price in Egypt, 2001–10

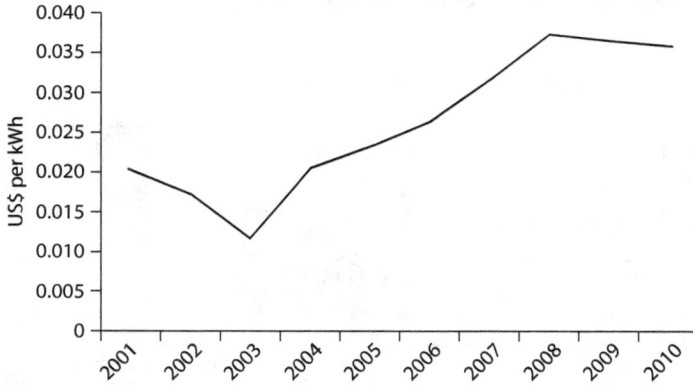

Source: EEHC.
Note: kWh = kilowatt-hour.

Figure 13A.6   Power Consumption Per Capita in Egypt, 1998–2008

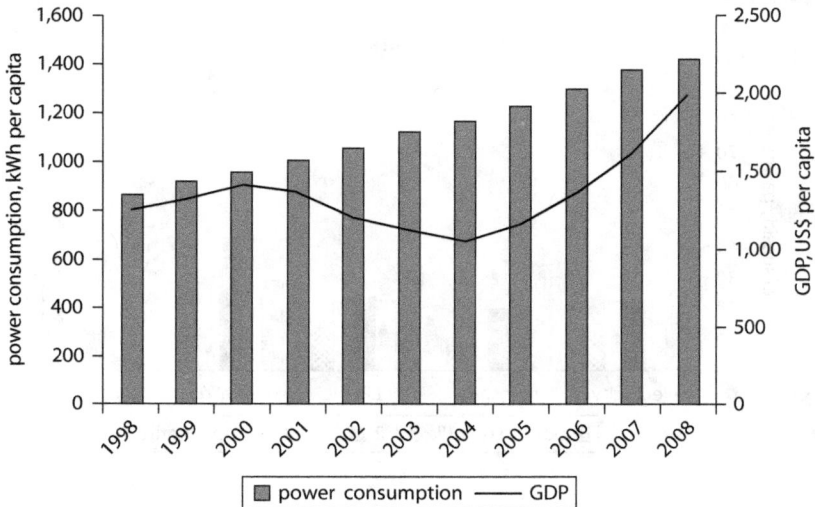

Source: World Bank, World Development Indicators.

Poverty Impact Evidence from Household Surveys in Egypt

Figure 13A.7   Electricity Block Tariffs in Egypt, 2010

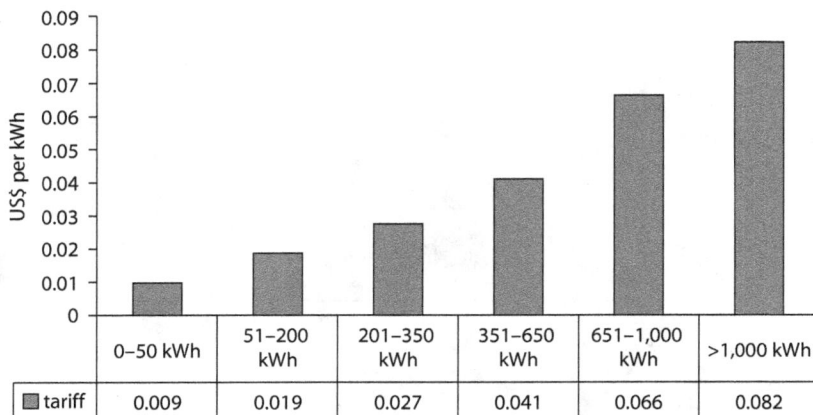

	0–50 kWh	51–200 kWh	201–350 kWh	351–650 kWh	651–1,000 kWh	>1,000 kWh
tariff	0.009	0.019	0.027	0.041	0.066	0.082

Source: EEHC.
Note: kWh = kilowatt-hour.

Figure 13A.8   Benefit Incidence of Energy Subsidies in Egypt, by Income Quintile, 2005

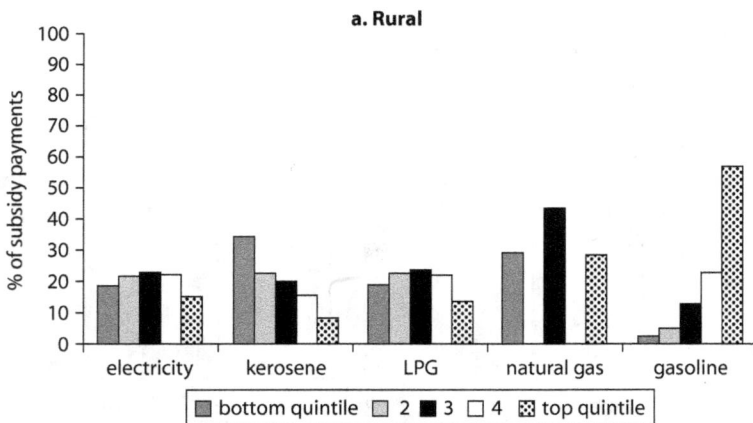

a. Rural

(continued next page)

Figure 13A.8   *(continued)*

b. Urban

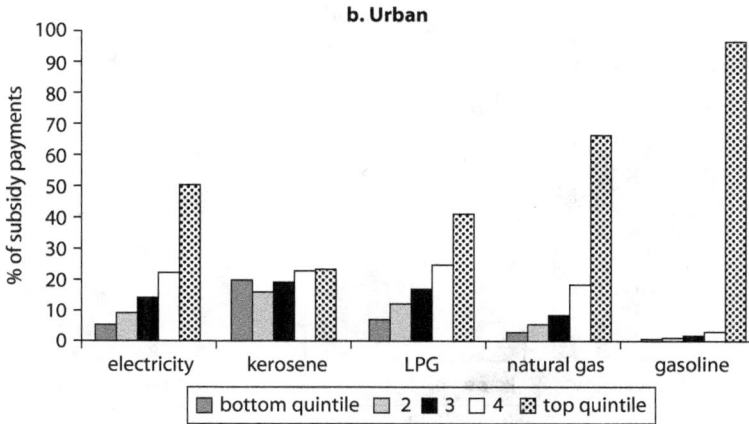

Source: World Bank 2009 data from 2005 household survey.
Note: LPG = liquefied petroleum gas.

Figure 13A.9   Household Energy Expenditure in Egypt, by Fuel and Income Quintile, 2005

a. Rural

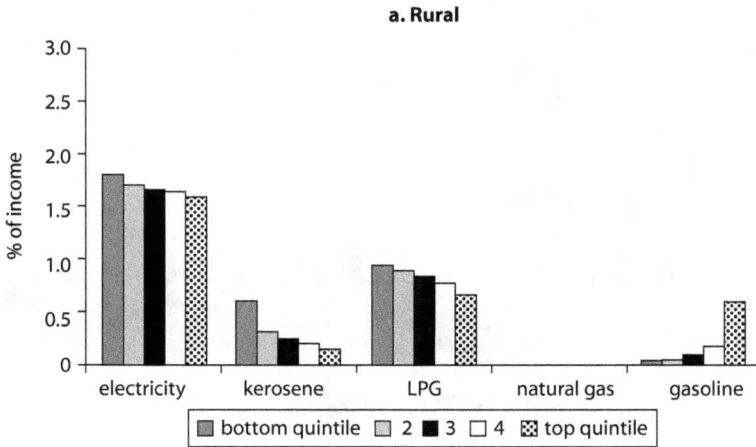

(continued next page)

Figure 13A.9   *(continued)*

b. Urban

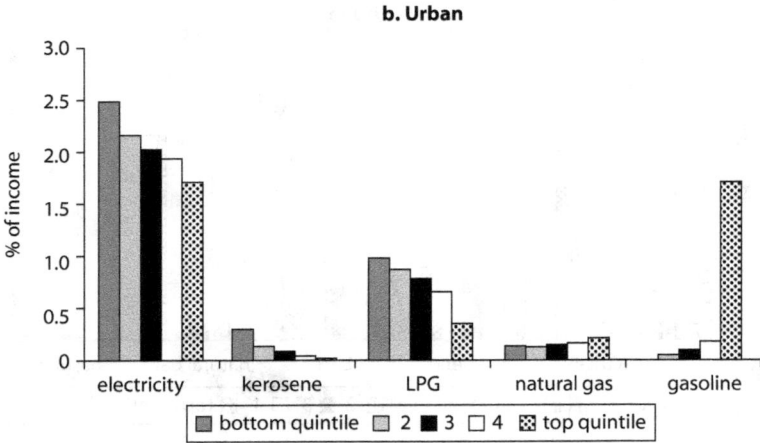

Source: World Bank 2009 data from 2005 household survey.
Note: LPG = liquefied petroleum gas.

Figure 13A.10   Household Energy Expenditure in Egypt, by Income Quintile, 2005–09

a. 2005

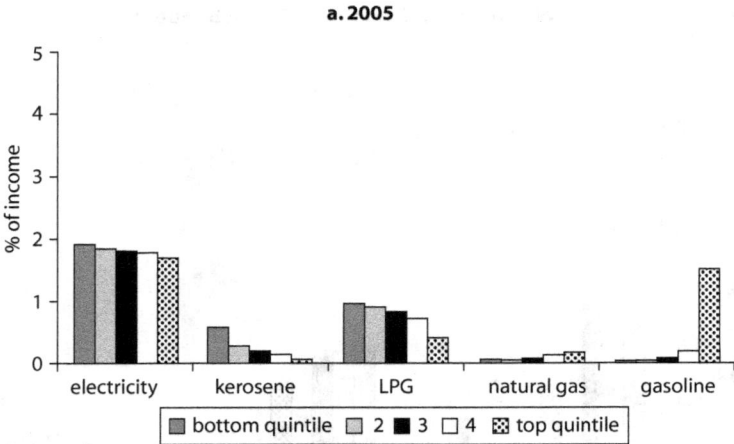

Source: World Bank 2009 data from 2005 household survey.
Note: LPG = liquefied petroleum gas.

(continued next page)

Figure 13A.10   *(continued)*

b. 2009

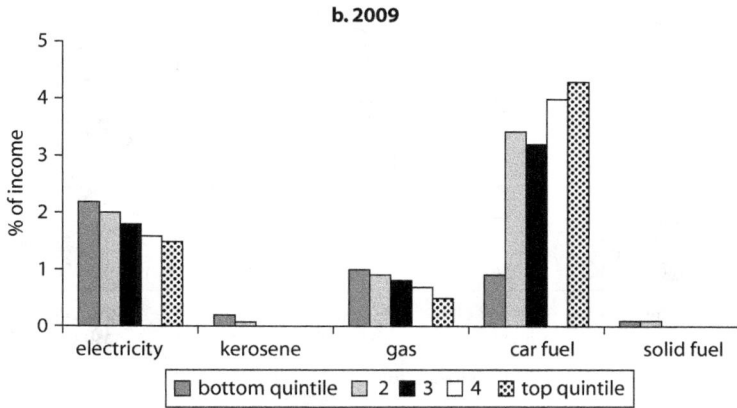

Source: Herrera 2010 data from 2009 household survey.
Note: "Car fuel" includes gasoline and diesel; "gas" = natural gas; "solid fuel" = biomass.

Figure 13A.11   Projected Welfare Impact of Removing Fuel Subsidies in Egypt, 2008–13

a. Rural

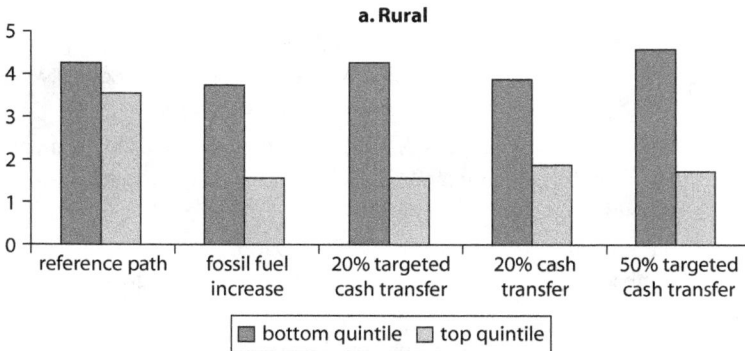

(continued next page)

Figure 13A.11 *(continued)*

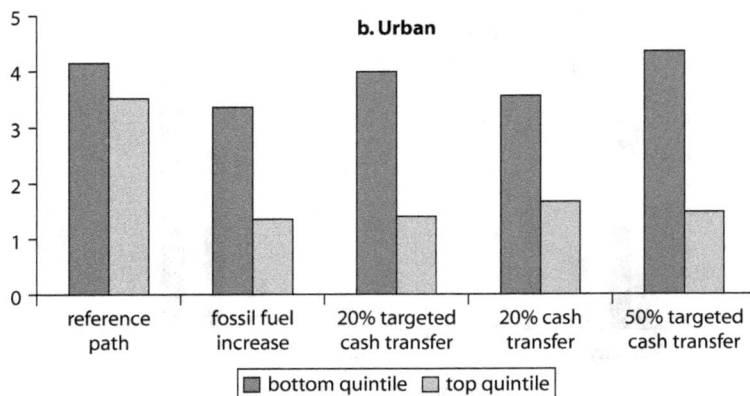

Source: Abouleinein, El-Laithy, and Kheir-El-Din 2009 and World Bank 2009 based on data from 2005 household survey.
Note: "Reference path" = May 2008 increase. "Fossil fuel increase" = scenarios without cash transfer. "20% targeted cash transfer" = redistribution of 20 percent of energy subsidy savings, targeted to the poorest two quintiles of the population. "20% cash transfer" = redistribution of 20 percent of energy subsidy savings to the population in general. "50% targeted cash transfer" = redistribution of 50% of energy subsidy savings, targeted to the poorest two quintiles of the population.

References

Abouleinein, Soheir, Heba El-Laithy, and Hanaa Kheir-El-Din. 2009. "The Impact of Phasing Out Subsidies of Petroleum Energy Products in Egypt." Working Paper 145, Egyptian Center for Economic Studies, Cairo.

CTF (Clean Technology Fund). 2009. "Clean Technology Fund: Investment Plan for Egypt." Working document CTF/TFC.2/7, CTF Trust Fund Committee, Climate Investment Funds, Washington, DC.

GIZ (German Agency for International Cooperation). n.d. International Fuel Prices database. GIZ (formerly GTZ), Bonn. http://www.gtz.de/en/themen/29957.htm.

Herrera, Santiago. 2010. "Subsidy Reform: Analysis and Perspectives." Remarks presented at the "Price Subsidies in Egypt: Alternatives for Reform" workshop, Egyptian Center for Economic Studies, Cairo, October 5.

IMF (International Monetary Fund). 2010. Retail domestic fuel prices data sheet, IMF, Washington, DC. http://www.imf.org/external/pubs/ft/spn/2010/data/spn1005.csv.

———. Various years. "Egypt: Article IV Consultations." IMF, Washington, DC.

World Bank. 2005. "Egypt: Toward a More Effective Social Policy: Subsidies and Social Safety Net." Poverty and Social Impact Analysis, Social and Economic Development Group, Middle East and North Africa Region, World Bank, Washington, DC.

———. 2009. "Arab Republic of Egypt: Consulting Services for an Energy Pricing Strategy." Final study report funded by the Energy Sector Management Assistance Program (ESMAP), World Bank, Washington, DC.

CHAPTER 14

Indonesia

Incentives to Energy Subsidy Reforms

Even though Indonesia has sizable resources of oil, gas, and coal reserves as well as large renewable (geothermal, hydropower, and solar) energy potential, it faces an enormous challenge in providing energy efficiently for the world's fourth-largest population, which is spread over a large archipelago of more than 6,000 inhabited islands. The government is looking to address looming energy shortages through two programs that will each add 10 gigawatts of electricity-generating capacity in 2011–15. Much of this new electricity capacity will be generated by coal-fired power plants. As a result, the proportion of total electricity generation using coal as a fuel will rise from 49 percent in 2010 to 55 percent in 2020, while the share accounted for by oil will fall from 23 percent to 9 percent. The government is also eager to make fuller use of Indonesia's large geothermal resources. A series of geothermal plants are planned, and by 2020 these are expected to generate 11 percent of the country's electricity, up from 3 percent in 2010.

Indonesia has a long history of energy subsidies. Suharto's new regime in 1967 continued fuel subsidies that were already in place. The subsidies increased significantly between 1998 and 2000, during the period around the Asian Financial Crisis and after depreciation of the Indonesian rupiah.

The appreciation of the rupiah in 2002–03 provided some subsidy relief for the government. However, the fuel price increases of 2003 brought about strong opposition to the Megawati government, which even forced it to undo the price increases and restore subsidies (Bulman, Fengler, and Ikhsan 2008).

As international oil prices almost doubled from 2003 to 2005, to about US$55 a barrel, the subsidy burden became larger for Indonesia and, at US$15 billion, made up nearly 5 percent of gross domestic product (GDP). Subsidies were reduced substantially as a result of reforms in 2005. However, with the rising oil price in 2008, the subsidy bill increased further, reaching a peak of 5.6 percent of GDP and 22 percent of the total government budget (see figure 14A.1). Fossil fuels represent the lion's share of energy subsidies. Subsidies to electricity remain sizable as well.

Reform Efforts

Fossil Fuels

Currently the state-owned energy company, Pertamina, operates eight refineries with a combined crude oil processing capacity of over 1 million barrels per day (although output is currently lower, at around 800,000 barrels per day). No new capacity has been added since 1994, which has led to an ever-increasing dependence on imported refined-petroleum products.

Until 2006, Indonesia was the world's largest producer of liquefied natural gas. Although production has suffered because of a lack of investment, it will continue to grow because two major blocks—the Tranche 4 gas field in West Java (with an estimated 80.48 billion cubic feet of reserves) and the Peciko Phase 7B well in Kalimantan (with an estimated 203.93 billion cubic feet of reserves)—came on stream in 2011.

The timing of subsidy reforms is summarized below:

- Domestic fuel prices were increased substantially in March and October 2005 as a result of the Yudhoyono government's decision to reduce the subsidy bill (see figure 14A.2), and consumption of diesel in the road sector declined by 10 percent (see figure 14A.3). The October increases were particularly sharp, raising the gasoline price by 88 percent, the diesel price by 105 percent, and the kerosene price by 186 percent in real terms (IMF 2007). As a result, average domestic fuel prices reached about 75 percent of the international level. This increase resulted in budgetary savings of about 0.5 percent of GDP in 2005 and an estimated 2.5 percent of GDP in 2006. Subsidies reappeared in 2007 as international oil prices increased further. Another domestic price increase followed in May 2008. In addition, the government stopped subsidizing large industrial electricity consumers.

- The 2010 and 2011 state budgets allowed the government to increase prices of fuel if the Indonesian crude price increased by more than 10 percent over US$80 per barrel. Each dollar increase above this threshold would increase the subsidy bill by Rp 2.7 trillion. However the government kept postponing the implementation of price increases, even when the prices were well above US$80 per barrel. The government was planning to reduce subsidies—partially by increasing prices for private car owners on January 1, 2011—but the implementation of this plan was postponed. According to the plan, subsidized fuel will be available only for public transportation and motorbikes.

- One of the most significant household fuel consumption reforms has been the government's program to reduce the use of kerosene, which is heavily subsidized, and replace it with liquefied petroleum gas (LPG). It is well known that although kerosene subsidies are intended for the poor, they are highly regressive and have caused the typical problems of smuggling, overuse, and fuel adulteration (World Bank 2007). The kerosene reduction program, which began in 2007, relies on the free distribution of LPG bottles and stoves as an incentive for households to switch from kerosene to LPG (IEA 2008). State oil and gas company Pertamina expected to complete this conversion program by mid-2010. So far, the program has been reasonably successful, increasing the domestic demand for LPG while reducing the demand for kerosene. Banten, Jakarta, West Java, Yogyakarta, and South Sumatra have been converted to LPG and are using relatively little kerosene. The conversion program is also causing an oversupply of kerosene in the approximate amount of 3 million barrels, which is now presenting export opportunities.

Electricity

The electric power industry in Indonesia is solely managed by Perusahaan Listrik Negara (PLN), a vertically integrated monopoly owned by the Ministry of State Owned Enterprises. PLN has various business units that carry out functions including generation, transmission, and distribution.

About three-quarters of the power produced in Indonesia is on the main island of Java. PLN controls about 85 percent of the total generation capacity, and the remainder is owned by large industries and mines (mainly for their own use) and by independent power producers that sell their output to PLN.

Electricity pricing, which is determined by the Ministry of Energy, is not cost-reflective. In 2005, the government removed PLN's fuel subsidy, but it did not allow the utility to pass higher fuel costs on through the retail electricity tariff. It is estimated that the average revenue received by PLN has been as low as US$0.06 per kilowatt-hour (kWh), while the average cost of production is US$0.12 per kWh (Vagliasindi and Besant-Jones, forthcoming). The difference is made up by the government (Ministry of Finance) through a direct subsidy to PLN. These subsidies make up about one-third of total energy subsidies, with two-thirds going to consumption fuels.

In line with the country's National Medium-Term Development Plan goal, electricity tariffs were increased in July 2010—the basic tariff by an average of 10 percent and the industrial tariff by an average of 10–15 percent. Small residential customers of lower voltage levels were shielded from price increases; their tariffs remained the same since 2004 (see table 14A.1). The structure of commercial tariffs was simplified. The impact of the increase in the electricity price was the reduction of subsidies by about 0.1 percent of GDP, which is still relatively small compared with the overall amount of energy subsidies. The government planned to increase electricity tariffs further (IMF 2010a).

Indonesia's per capita power consumption significantly increased (see figure 14A.6) and surpassed other countries such as India and Nigeria, which have significant energy subsidies (World Bank 2007).

Poverty Alleviation Measures

Evidence from Household Surveys

The pattern of consumption in terms of energy use is similar in rural and urban areas, with the notable exception of biomass, which is used mainly in rural areas by the bottom-income quintiles (see figure 14A.7).

Electricity is the most burdensome source of energy for households, but whereas in rural areas the percentage of household expenditure generally increases with income (mainly due to the higher use), in urban areas it is neither progressive nor regressive (see figure 14A.8). Kerosene accounts for a rather high share of the household budget, particularly in urban areas and among the poor. Biomass displays a regressive expenditure pattern in both urban and rural areas. On the other hand, petroleum products and LPG expenditures are progressive for both urban and rural households. There might have been changes in recent years because of the implementation of the kerosene-to-LPG program.

Subsidies in Indonesia are particularly badly targeted and are applied to transportation fuels as well as to fuels used for household purposes. According to a May 2008 statement by the Indonesian Ministry of Economic Affairs, the top 40 percent of households in terms of income benefit from 70 percent of the subsidies while the bottom 40 percent benefit from only 15 percent of the subsidies. Indonesia's subsidized fuel prices are among Asia's cheapest—around 30 percent of the world price (IEA 2008). Gasoline and diesel each account for roughly one-quarter of all subsidy spending (Bulman, Fengler, and Ikhsan 2008). A World Bank study showed that the richest 10 percent benefit five times more from gasoline subsidies than the poorest 10 percent (World Bank 2007). As in other countries with blanket subsidies, because of cross-border price differentials, there is room for benefiting from fuel smuggling. The government loss from fuel smuggling in 2005 was estimated at around US$850 million (IEA 2008).

Del Granado, Coady, and Gillingham (2010) consider the welfare impact of a $0.25 per liter increase in fuel prices in the case of Indonesia (see figure 14A.9), considering only the direct impact on the consumption of fuels for cooking, heating, lighting, and private transport. The impact results in the loss of slightly below 9 percent of real income, mostly due to the reduction in kerosene and electricity consumption (by 4.1 percent and 3.8 percent, respectively).

Yusuf (2008) used a computable general equilibrium model to simulate the economic and distributional impact of the package of subsidy removal implemented by the Indonesian authorities in October 2005, which entailed the increases in the price of gasoline by 87.5 percent, diesel by 104.7 percent, and kerosene by 185.7 percent. The 2005 package is simulated through the following scenarios (Yusuf 2008):

- *Scenarios 1.A and 1.B* represent the 2005 reform package, respectively without or with the increase in the price of kerosene.
- *Scenario 2.A* represents the 2005 reform package, considered together with the introduction of compensation through an unconditional cash transfer (UCT) to the poor.
- *Scenarios 2.B and 2.C* represent the introduction of targeted cash transfers, each with a different degree of effectiveness (100 percent and

75 percent, respectively). (Because there was no consensus on the degree of effectiveness of the UCT scheme, two alternative assumptions about its effectiveness were made—100 percent effective and 75 percent effective—the latter scenario assuming that the amount of cash given to every targeted group of households is reduced by 25 percent.)

- *Scenarios 3.A and 3.B* represent the introduction of a conditional subsidy to targeted households for spending on education and health (in the same amount of the UCT), respectively with and without the 2005 reform package.

The distributional impacts of the alternative scenarios shows interesting results (see figure 14A.10). Incidence of both urban and rural poverty is significantly lower where the subsidy removal does not include kerosene (Scenario 1.A relative to Scenario 1.B), supporting the evidence reported in the previous section that among the petroleum fuels, kerosene is the most "progressive." Overall national poverty decreases for all UCT schemes, but urban poverty is shown to increase even with a 100 percent effective UCT scheme (Scenarios 2.A and 2.B). Adjusting somewhat the scheme—providing more support to the urban poor—is shown to prevent the increase in urban poverty while not negatively affecting poverty in rural areas. In contrast with the UCT schemes, subsidizing education and health expenditure, when combined with the 2005 package, would increase both urban and rural poverty by 0.9 and 0.35 percent, respectively (Scenarios 2.A and 2.B). This result may be interpreted as investment in human capital through a longer-term strategy rather than a short-term measure. Hence it is less effective in minimizing the impact of energy pricing reforms.

Social Safety Nets

Indonesia has been particularly successful at designing targeted cash transfers that were passed on simultaneously with fuel price increases in 2005. The UCT program is the largest such program in the world—covering 19.2 million households, or one-third of the Indonesian population. The program was introduced after the October 2005 price increases.

Before execution of the transfers, each household was given a proxy means test. Recipients were issued smart cards (with instructions printed on the back of the cards), and transfers were delivered through the post office system. The program delivered benefits of US$30 per quarter, significantly more than the increase in energy costs. This served to increase the level of assistance for the poor and to buy their acquiescence to the fuel price increases. At the same time, by covering the bottom two income quintiles (40 percent) of the population—more than the targeted bottom 28 percent—the program also helped prevent those on the verge from falling into poverty (Bacon and Kojima 2006). In addition to transferring cash to the lowest-income households, the government also used the savings from the decreased fuel subsidies to finance programs in education, rural development, and health.

Key Lessons Learned

The earlier stages of the subsidy reform in Indonesia are largely regarded as successful, particularly given the size of the program and previous episodes of unrest. The speed with which the UCT was designed and implemented meant that some leakage, targeting errors, and logistical difficulties were inevitable. However, the government responded quickly to reports of irregularities and, in spite of the challenges, the program proved largely successful in reaching the poor: the poorest decile received 21 percent of the benefits, while deciles 2, 3, and 4 captured 40 percent. In the absence of compensation, the fuel price hikes would have led to an estimated 5 percent rise in the poverty headcount index. In recent years, however, there has been only limited progress in the area of subsidy reforms because domestic prices of fuels and electricity remain mostly fixed by the government.

Indonesia has a history of violent protests against attempts to implement fuel price increases—for instance, when President Megawati was forced to roll back prices in 2003. However, an important lesson that emerges from the case of Indonesia is that decisive leadership and government popularity, along with appropriate compensation measures and an effective information campaign, work to counteract citizen disenchantment and prevent any public discontent. Current President Yudhoyono's credibility helped to successfully increase fuel prices, the savings from which he then directed toward the UCT program in 2006. Additionally, before introducing UCT, the government initiated an extensive nationwide information campaign about the benefits of the program, which helped citizens learn about the value of the program and prevented possible unrest.

Another important lesson from the case of Indonesia is that there is no case for universal subsidization of kerosene and even less of one for subsidization of gasoline and diesel, which find largely commercial uses. The increased use of LPG—which is a cleaner, safer, and better-performing cooking and lighting fuel—should be praised. In addition to LPG, the government may also want to promote the use of coal briquettes or similar alternatives to kerosene where LPG may not be available (World Bank 2007). By shifting away from subsidized fuels and below-market electricity tariffs, Indonesia will make room for the renewable energy technologies that it is keen to scale up. Further regulatory improvements to invite investment in the energy sector and renewable energy feed-in tariffs would facilitate the process.

Annex 14.1 Indonesia Case Study Figures

INCOME LEVEL: Lower-middle income
REGION: East Asia and Pacific
ENERGY NET IMPORTER/EXPORTER: Net exporter
SUBSIDIES: Kerosene, gasoline, diesel, electricity
PHASING OUT SUBSIDIES: Partly successful

Fiscal Burden of Energy Subsidy in Indonesia

Figure 14A.1   Explicit Budgetary Energy Subsidies in Indonesia, 1997–2010

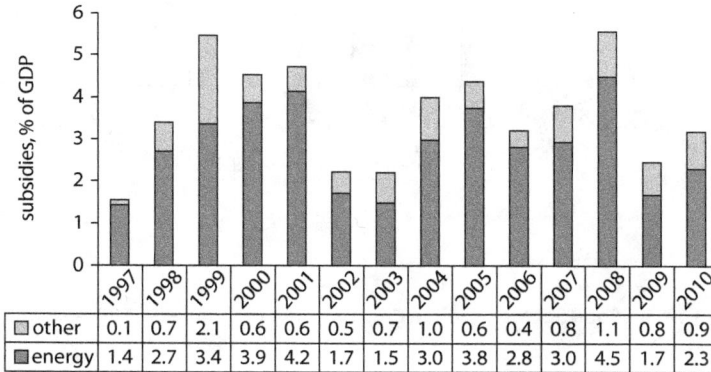

	1997	1998	1999	2000	2001	2002	2003	2004	2005	2006	2007	2008	2009	2010
□ other	0.1	0.7	2.1	0.6	0.6	0.5	0.7	1.0	0.6	0.4	0.8	1.1	0.8	0.9
■ energy	1.4	2.7	3.4	3.9	4.2	1.7	1.5	3.0	3.8	2.8	3.0	4.5	1.7	2.3

Sources: IMF staff reports and Ministry of Finance, various years.

Fuel Prices and Road Sector Consumption in Indonesia

Figure 14A.2   Domestic Retail Fuel Prices in Indonesia, 2002–10

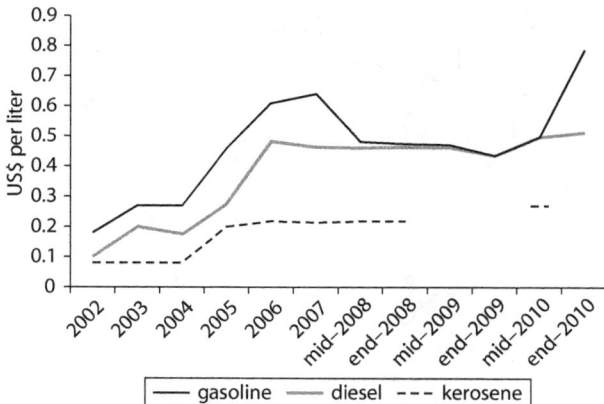

gasoline —— diesel --- kerosene

Source: Elaboration of data from GIZ n.d.; IMF 2010b; and additional data from individual country information.

Figure 14A.3   Road Sector Diesel Consumption in Indonesia, 1998–2008

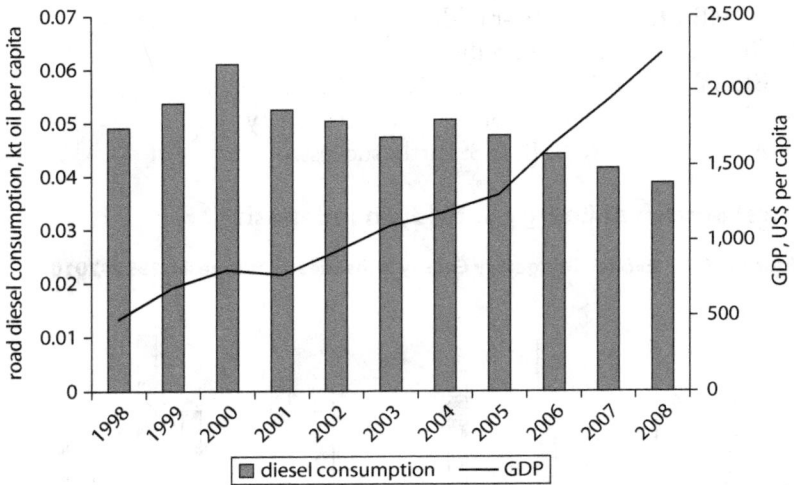

Source: World Bank, World Development Indicators.

Figure 14A.4   Road Sector Gasoline Consumption in Indonesia, 1998–2008

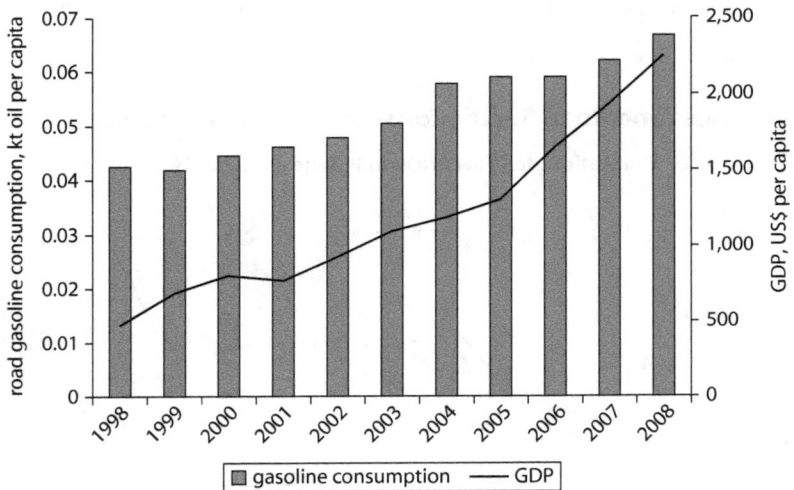

Source: World Bank, World Development Indicators.

Electricity Price and Power Consumption in Indonesia

Figure 14A.5   Electricity Price in Indonesia, 2000–10

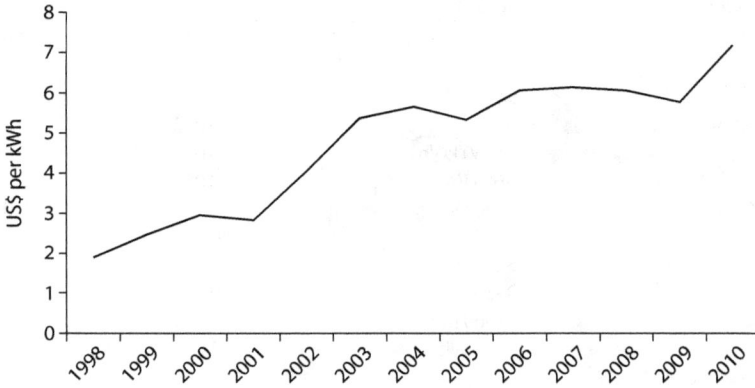

Source: PLN.
Note: kWh = Kilowatt-hour.

Figure 14A.6   Power Consumption Per Capita in Indonesia, 1998–2008

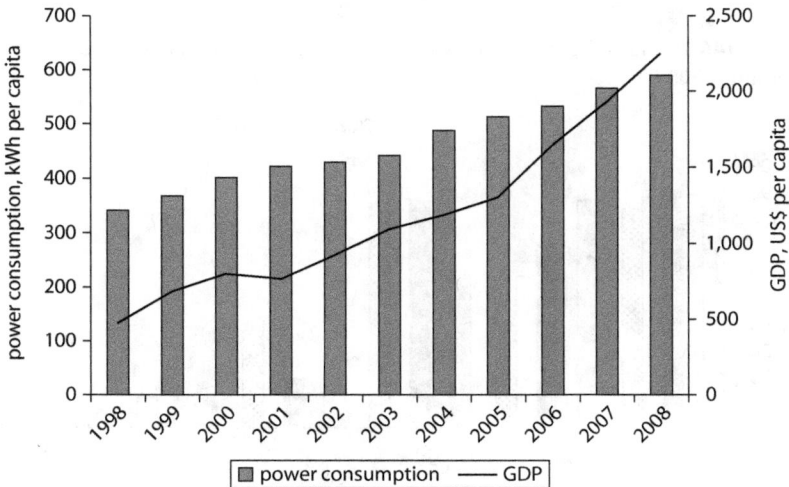

Source: World Bank, World Development Indicators.
Note: kWh = Kilowatt-hour.

Poverty Impact Evidence from Household Surveys in Indonesia

Table 14A.1   Electricity Block Tariffs in Indonesia, 2004 and 2011

Power limit	Block (kWh)	2004 (US$)	2011 (US$)
450 VA	0–30 kWh	0.019	0.019
	31–60 kWh	0.039	0.039
	>60 kWh	0.054	0.054
900 VA	0–30 kWh	0.030	0.030
	31–60 kWh	0.049	0.049
	>60 kWh	0.054	0.054
1,300 VA	0–30 kWh	0.042	0.0868
	31–60 kWh	0.049	
	>60 kWh	0.054	
2,200 VA	0–30 kWh	0.043	0.0873
	31–60 kWh	0.049	
	>60 kWh	0.054	
2,200–6,600 VA		0.098	0.098
>6,600 VA		H1 x 0.098	H1 x 0.098
		H2 x 0.152	H2 x 0.152

Source: PLN Tariff Order, 2004 and 2011.
Note: kWh = kilowatt-hour; VA = volt-ampere; H1 = first block; H2 = second block.

Figure 14A.7   Household Energy Use in Indonesia, by Source and by Income Quintile, 2005

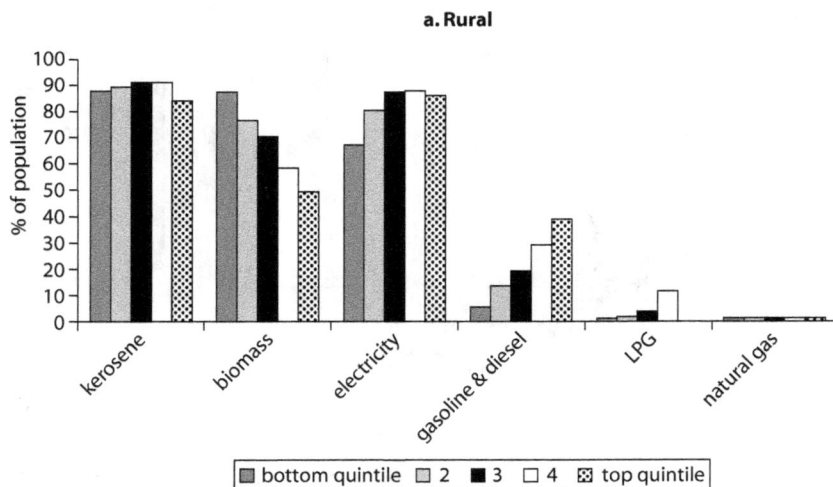

(continued next page)

Figure 14A.7   *(continued)*

b. Urban

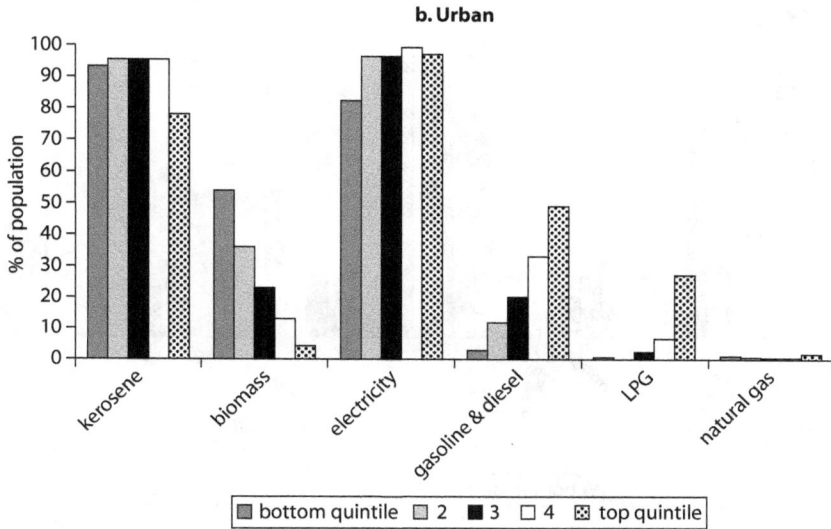

Source: Rand Corp. 2005, from household survey data.
Note: LPG = liquefied petroleum gas.

Figure 14A.8   Household Energy Expenditure in Indonesia, by Income Quintile, 2005

a. Rural

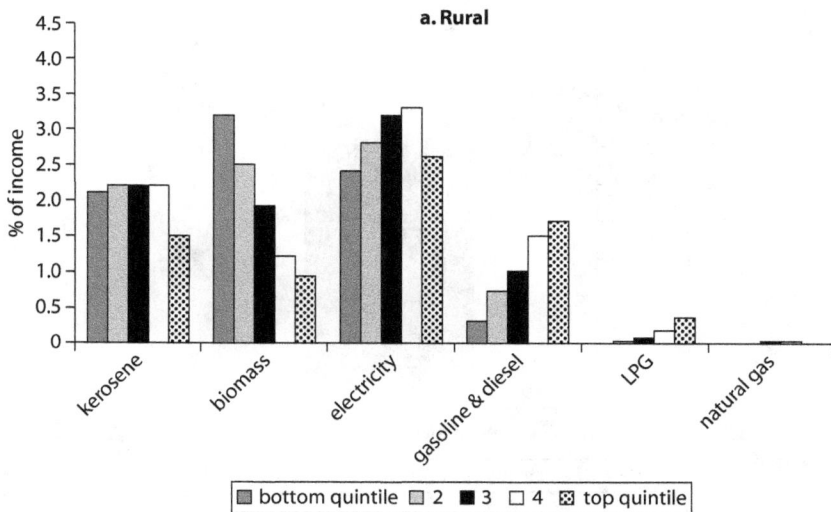

(continued next page)

Figure 14A.8 *(continued)*

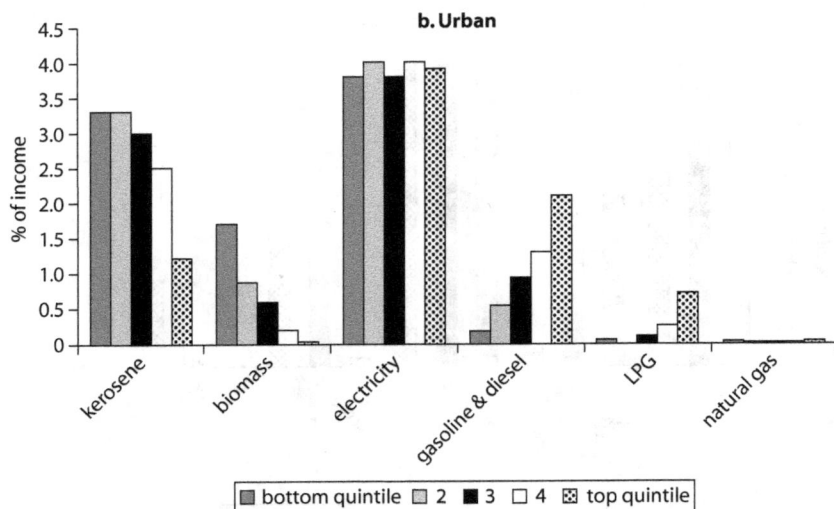

b. Urban

Source: Rand Corp. 2005, from household survey data.
Note: LPG = liquefied petroleum gas.

Figure 14A.9   Welfare Impact of Removing Fuel Subsidies in Indonesia, 2005

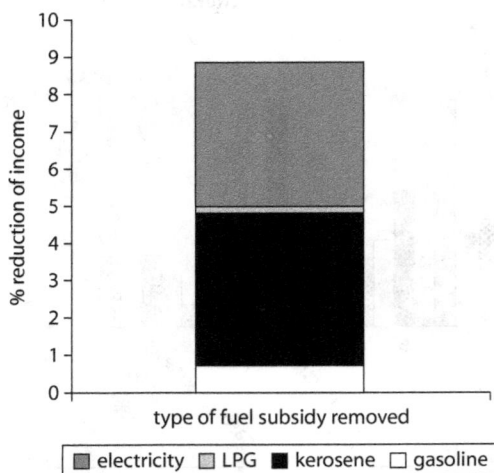

Source: Del Granado, Coady, and Gillingham 2010.
Note: LPG = liquefied petroleum gas.

Figure 14A.10   Welfare Impact of Removing Fuel Subsidies in Indonesia

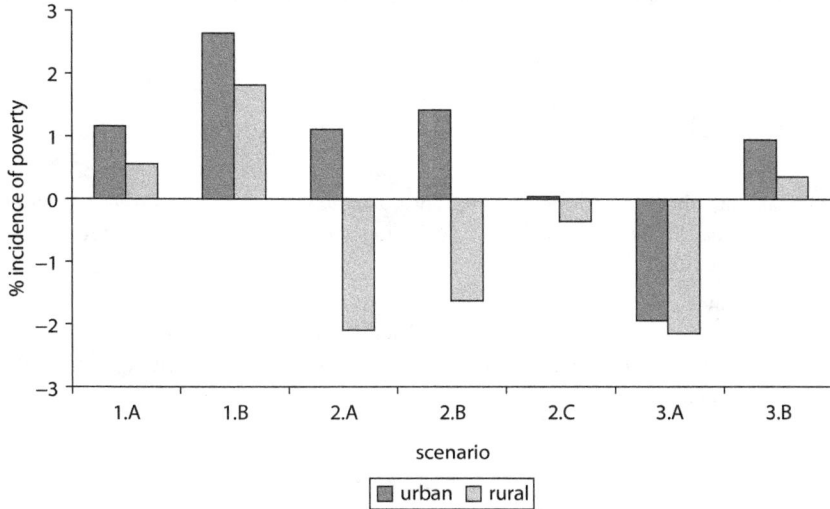

Source: Yusuf 2008.
Note: Scenarios 1.A, 1.B = impact of 2005 reform package, respectively without or with the increase in kerosene price. *Scenario 2.A* = 2005 reform package plus UCT. *Scenarios 2.B, 2.C* = targeted UCT, assuming respectively 100 percent and 75 percent effectiveness. *Scenarios 3.A, 3.B* = conditional subsidy to targeted households for education and health spending (in same amount as UCT), respectively with and without the 2005 reform package. UCT = unconditional cash transfer.

References

Bacon, Robert, and Masami Kojima. 2006. "Coping with Higher Oil Prices." Report 323/06, Energy Sector Management Assistance Program (ESMAP) and World Bank, Washington, DC.

Bulman, Tim, Wolfgang Fengler, and Mohamad Ikhsan. 2008. "Indonesia's Oil Subsidy Opportunity." *Far Eastern Economic Review* 171 (5): 14–18.

Del Granado, J., D. Coady, and R. Gillingham. 2010. "The Unequal Benefits of Fuel Subsidies: A Review of Evidence for Developing Countries." Working Paper 10/202, International Monetary Fund, Washington, DC.

GIZ (German Agency for International Cooperation). n.d. International Fuel Prices database. GIZ (formerly GTZ), Bonn. http://www.gtz.de/en/themen/ 29957.htm.

IEA (International Energy Agency). 2008. *Energy Policy Review of Indonesia.* Paris: Organisation for Economic Co-operation and Development/IEA.

IMF (International Monetary Fund). 2007. "Indonesia Country Report." IMF, Washington, DC.

———. 2010a. "Indonesia: Article IV Consultation." IMF, Washington, DC.

————. 2010b. Retail domestic fuel prices data sheet, IMF, Washington, DC. http://www.imf.org/external/pubs/ft/spn/2010/data/spn1005.csv.

Rand Corp. 2005. SUSENAS (National Socio-economic Survey). http://www .rand.org/labor/bps/susenas.html.

World Bank. 2007. "Toward an Efficient Fuel Products Market in Indonesia: Achieving an Equitable and Sustainable Policy." World Bank, Washington, DC.

Vagliasindi, Maria, and John Besant-Jones. Forthcoming. *Power Market Structure: Revisiting Policy Options*. Directions in Development Series. Washington, DC: World Bank.

Yusuf, Anshory. 2008. "The Distributional Impact of Environmental Policy: The Case of Carbon Tax and Energy Pricing Reform in Indonesia." EEPSEA Research Report 2008-RR1, Economy and Environment Program for Southeast Asia (EEPSEA), Singapore.

CHAPTER 15

Islamic Republic of Iran

Incentives to Energy Subsidy Reforms

Fuel Subsidies

With oil reserves at 137.6 billion barrels at the end of 2009 (10.3 percent of the world's total and second only to Saudi Arabia), the Islamic Republic of Iran is the second-largest oil exporter in the world. Its natural gas reserves stood at 1,046 trillion cubic feet in 2008—15.8 percent of the world's total and second only to the Russian Federation. The industry is also of central importance to the health of the Iranian economy. The oil and gas sector accounted for around 18 percent of gross domestic product (GDP) in 2009.

Excessive gasoline consumption has turned the Islamic Republic of Iran into a net gasoline importer, with a negative impact on its fiscal balance (IMF 2009). The predictable results were rising subsidy bills for the state (see figure 15A.1). Most of this burden was carried as an implicit subsidy to domestic energy consumers, with the price of diesel fuel, for example, set at the equivalent of US$0.02 per liter, and gasoline selling for less than bottled water. The country's fuel prices were among the lowest in the world, only ahead of Venezuela's. This has encouraged excessive use of energy (both in per capita terms and per unit of GDP)—reflected by the Islamic Republic of Iran's move from being one of the least-energy-intensive users in the world in 1980 to one of the most-intensive users in 2009. The country's energy intensity was eight times more than that of any European country.

Low fuel prices have not only caused consumption inefficiency in the country but also have presented an opportunity for fuel smuggling in neighboring countries such as Afghanistan, Iraq, Pakistan, and Turkey. The price of gasoline in Turkey used to be 20 times higher than in the Islamic Republic of Iran, and diesel more than 50 times higher. According to International Monetary Fund estimates, implicit oil subsidies cost the government more than 10 percent of GDP annually (see figure 15A.2). If natural gas and electricity subsidies are also included, the cost of subsidies averaged more than 20 percent of GDP from 2006 to 2009.

Electricity Subsidies

The electricity tariff structure in the Islamic Republic of Iran is complex. There are six separate tariff classes: for residential customers, public services, agricultural use, industrial use, street lighting, and the commercial sector. Each customer tariff category has time-of-use and seasonal features and various discounts and premiums, and many have capacity and energy charges and a number of tariff blocks (World Bank 2007).

The government heavily subsidizes end-user tariffs by providing fuel to the power plants at prices well below their economic costs. A simplified calculation for fiscal year (FY) 2007 demonstrates the extent of the subsidies in the sector: Assuming that the new fuel prices introduced in 2007 were cost-reflective, the total cost of electricity supply in FY 2007 is estimated to have been about US$12.1 billion. The Islamic Republic of Iran's power generation, transmission, and distribution management company, Tavanir, collected only US$2.7 billion (or 22 percent) of this cost from consumers at the prevailing tariffs (an average of Rls 164.68 per kilowatt-hour [kWh], or about US$0.018 per kWh). The rest was a subsidy, borne either by the government budget or by fuel suppliers.

Efficiency also requires that tariffs reflect the costs of supply. The average consumer tariff in FY 2007—Rls 165 per kWh (US$0.018 per kWh)—masks significant differences between consumer categories. On average,

- Agriculture paid Rls 20 per kWh (US$0.002)
- Households paid Rls 130 per kWh (US$0.014)
- The public sector paid Rls 170 per kWh (US$0.019)

- Industry paid Rls 200 per kWh (US$0.022)
- The commercial sector paid Rls 510 per kWh (US$0.056).

The commercial sector was the only customer class that covered its average supply cost in FY 2007, net of fuel subsidies (World Bank 2007).

Another big challenge that the Islamic Republic of Iran has been facing in its energy sector is meeting the rising electricity demand, which has surpassed its GDP growth. Electricity consumption has been increasing substantially over time (see figure 15A.7) to the point that electricity consumption per capita was three times the world average. Low end-user electricity prices have certainly been one of the factors in higher electricity demand, suggesting that energy use is largely inefficient. After the 2010 reforms, energy consumption reportedly declined by up to 20 percent for some fuels (IMF 2011). The quadrupling of the price of gasoline in the country is estimated to have reduced daily gasoline consumption from about 66 million liters to about 54 million liters in the first weeks following the reforms. The household consumption of natural gas declined from 200 million cubic meters to 180 million.[1] At the same time, the use of public transportation increased sharply.

Reform Efforts

Fossil Fuels

Tariffs for fossil fuels were substantially increased in 2007. Prices for diesel increased from Rls 58.6 per liter (US$0.0063) to Rls 4,477 per liter (US$0.484); and for heavy fuel oil from Rls 30.8 per liter (US$0.0033) to Rls 2,803 per liter (US$0.303). The new gasoline prices, at slightly above US$2 per thousand British thermal units (mBtu), compared with prices in the United States and Europe then of between US$6 and US$7 per mBtu, respectively. In June 2007, the country started addressing the problem of excessive consumption and subsidies by making fuel rationing mandatory, which has been reflected in reduced gasoline consumption (see figure 15A.5). The price of the rationed fuel was US$0.10 per liter, and if drivers exceeded their quota, they were obliged to buy it at a price of US$0.40 per liter. The rationed amounts have been administered by smart cards.

The parliament started to discuss the reform bill to remove subsidies in late 2008. However, it took a couple of years for the bill to become law because of the opposition in parliament. The international global financial crisis and United Nations sanctions against the Islamic Republic of Iran speeded up the process. The Subsidy Reform Law was approved by the Islamic Assembly on January 5, 2010, and confirmed by the Guardian Council on January 13, 2010. The start of the reforms was announced by President Ahmadinejad in December 2010. The bill introduced some drastic measures that would reduce or eliminate subsidies on a number of commodities and services, including gasoline, diesel fuel, fuel oil, kerosene, liquefied petroleum gas, other oil products, natural gas, and electricity. The reforms were to be implemented in a number of steps over a five-year period (2010–15). Energy carriers would have to gradually adjust the prices of petroleum products up to a level of not less than 90 percent of Persian Gulf free-on-board prices. The price of natural gas would be adjusted up to a level of not less than 75 percent of the average export price of natural gas. The price of electricity would be adjusted until reaching full cost recovery.

The increase in prices of energy products, public transport, wheat, and bread adopted in December 2010 are estimated to have removed close to US$60 billion (about 15 percent of GDP) in annual implicit subsidies to products. At the same time, the redistribution of the revenues arising from the price increases to households as cash transfers has been effective in reducing inequalities, improving living standards, and supporting domestic demand in the economy. The energy price increases are already leading to a decline in excessive domestic energy consumption and related energy waste. Although the subsidy reform is expected to result in a transitory slowdown in economic growth and a temporary increase in the inflation rate, it should considerably improve the Islamic Republic of Iran's medium-term outlook by rationalizing domestic energy use, increasing export revenues, strengthening overall competitiveness, and bringing the country's economic activity closer to its full potential.

The authorities have been successful in containing the initial impact of the energy price increases on inflation. Despite the large price increases of up to 20 times, consumer price inflation has only increased from 10.1 percent in December 2010 to 14.2 percent at the end of May 2011. Maintaining macroeconomic stability in the near term through coordinated and adequately tightened monetary and fiscal policies is essential to preserve the benefits of the subsidy reform. Equally challenging will be the restructuring of enterprises through the adoption of more energy-efficient technologies and the broader reorientation of the economy toward less energy-intensive products, services, and production technologies (IMF 2011). The results of a simulation of the increase in the price of gasoline indicate that the quadrupling of the price would allow reduction of gasoline consumption from 66 million liters to 54 million liters. The price adjustment would bring an additional Rls 81 billion in annual sales from both the domestic market and from exports (Guillaume, Zytek, and Farzin 2011).

Electricity

In 2003, the government started to engage in a major reform program of the electricity sector, including unbundling of the industry into a number of vertically integrated regional electricity companies (RECs) and the creation of a mandatory spot market. A separate company (the Iranian Grid Management Company) was charged with operating the spot market and the transmission network (owned by the RECs). An Electricity Market Regulatory Board within the Ministry of Energy was set up to regulate the spot market. All the state-owned companies in the sector were put under the umbrella of a holding company, Tavanir, governed by the Ministry of Energy.

The price increase implemented in 2007 was substantial in real terms (see figure 15A.6). For natural gas, the price rose from Rls 49.26 per cubic meter (US$0.0053, or about US$0.13 per mmBtu) to Rls 690 (US$0.0746). In the summer of 2008, Tavanir had to resort to regular load shedding, an indication that the sector reform program as implemented so far had been unsuccessful in curing the sector's problems.

The use of the multitier tariffs on electricity, natural gas, and water played an important role in moderating the impact of the price increases on small users, mostly the poor, and accounting for regional disparities in availability of different heating fuels. Unit tariffs on electricity, natural gas, and water use were set using escalating schedules (see, for example, figure 15A.8). The cost of the first 100 kWh of electricity use was set at a low price of just Rls 270 (about US$0.027). However, unit prices were set to rise rapidly, all the way to Rls 2,100 for use in excess of 600 kWh. Large household consumers were charged prices marginally higher than in international markets. The tariff schedules were further differentiated by region. For example, prices were set at lower rates in hot regions with relatively higher air-conditioning demand. Tariff schedules for natural gas and water were similarly differentiated by quantity used and region. In areas where natural gas was not available, heating costs were to remain closely monitored and regulated, and lower-priced rationed kerosene (at Rls 1,000 per liter) and lower electricity rates would be provided to ensure affordability of heating.

Poverty Alleviation Measures

Social Safety Nets

A review of the Islamic Republic of Iran's social assistance program preceding the 2010 reform shows that the system was highly fragmented and lacked appropriate targeting, monitoring, and impact evaluation systems. Lack of policy coordination was observed at all levels of the social protection system. The State Welfare Organization is the governmental agency that provides social services and assistance. The country has nearly 100 social assistance programs to help the needy, which are managed by some 30 public and quasi-public institutions. Many of these programs have overlapping target beneficiaries, and their coverage, targeting, and administration need to be reviewed (World Bank 2007).

Before the 2010 price reforms, price subsidies and rations represented the major safety net for the country's poor. Subsidies on a variety of commodities were used as the main social safety net. There were price controls on 39 different consumption items, with energy featuring as the most heavily subsidized group of commodities. The top income deciles were benefiting the most from electricity subsidies (see figure 15A.9), but reforms were made difficult by the substantially higher proportion of expenditure on utilities as a percentage of income by the poor (see figure 15A.10).

The smart card scheme associated with fuel rationing in 2007 was characterized by administrative failings and poor organization: cards arrived late, monthly rations and above-quota prices were not announced, there was no provision for the replacement of faulty cards, card swapping was not controlled, and so on. In short, the policy failed to affect the long-term, liberal consumption habits.

To protect the population from dramatic price increases, the 2010 reforms included a compensatory scheme in the form of cash transfers. Initially, the government was planning to provide compensation only to the poorest households. However, it became clear that it would be difficult to correctly identify the recipients. Thus, it was decided to provide universal compensation. The rich were asked to refrain from applying for the compensation. The application process was made as simple as possible, and it does not include income verification. Upon the announcement of the start of the reforms, about 80 percent of the population was given access to the bank deposits that contained compensatory payments.

The subsidy reform law stipulated that, out of the revenue arising from the price increase, about US$30 billion was to be redistributed in the first year to households in the form of freely usable cash; US$15–18 billion to enterprises to finance their restructuring to reduce energy intensity; and around US$10–12 billion to the government to allow it to pay for higher energy bills and improve energy efficiency in the public sector (IMF 2011; Guillaume, Zytek, and Farzin 2011). The compensation represents a large share of the budget of poor households and should alleviate poverty for some of them. A compensation of US$40 per person per month represents more than 50 percent of the labor income for many rural households. Direct compensation will encourage households to reduce their consumption of energy and will provide them with funds to buy other goods and services.

The distribution of cash transfers to households is estimated to have reduced poverty incidence from 12 percent to 2 percent, on the basis of a $2-per-day poverty line. Moreover, regional disparities in poverty have sharply declined. The distribution of cash transfers markedly improved income distribution. As a result, the Gini coefficient is estimated to have fallen from 0.40–0.45 before the reform to 0.37 after the implementation of the subsidy reform.[2] However, there are concerns about the sustainability of the program because of its wide coverage.

Key Lessons Learned

The Islamic Republic of Iran is one of the first oil exporting countries to have implemented drastic subsidy reforms. Thus, other countries will look to its experience when introducing changes in their own economies.

It is still too early to make conclusions about the effects of the country's subsidy reforms. However, it can be said that the government succeeded in initiating reforms in a smooth way and avoiding any serious public unrest. There are also reports of sharp declines in the consumption of fuels that followed the implementation of the program.

The key elements of the reforms that helped to avoid serious problems at the outset of the changes were these:

- Simplicity of the compensatory scheme
- A well-designed, broad public campaign
- Readiness of the banking system to facilitate massive cash transfers
- Government policies that helped to avoid dramatic increases in the inflation rate.

Phasing out subsidies in the Islamic Republic of Iran is not likely to be a near-term accomplishment, and the future success of the reforms will depend to a large extent on the macroeconomic policies of the Iranian government and the willingness and ability of the enterprises to restructure.

Annex 15.1 Islamic Republic of Iran Case Study Figures

INCOME LEVEL: Lower-middle income
REGION: Middle East and North Africa
ENERGY NET IMPORTER/EXPORTER: Net exporter
SUBSIDIES: Electricity, gasoline, diesel
PHASING OUT SUBSIDIES: Ongoing

Fiscal Burden of Energy Subsidy in the Islamic Republic of Iran

Figure 15A.1   Explicit Budgetary Energy Subsidies in the Islamic Republic of Iran, 2002–10

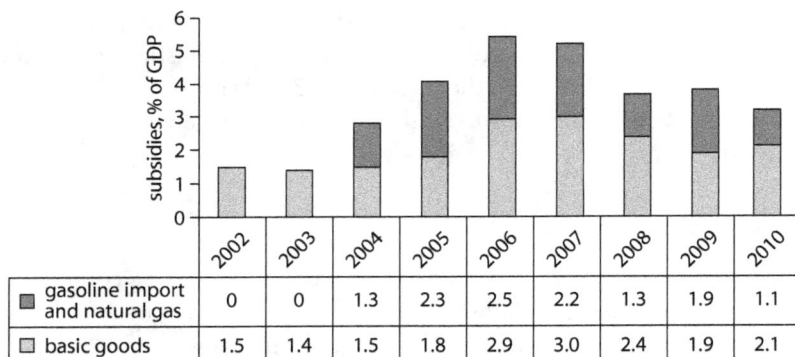

	2002	2003	2004	2005	2006	2007	2008	2009	2010
gasoline import and natural gas	0	0	1.3	2.3	2.5	2.2	1.3	1.9	1.1
basic goods	1.5	1.4	1.5	1.8	2.9	3.0	2.4	1.9	2.1

Source: IMF.

Figure 15A.2   Implicit Energy Subsidies in the Islamic Republic of Iran, 2006–09

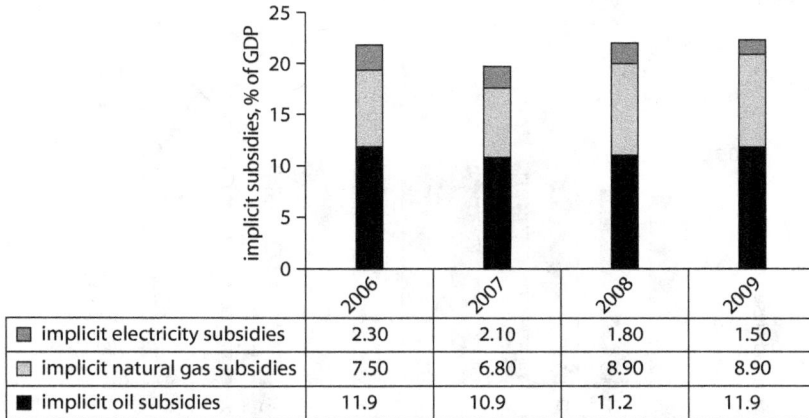

	2006	2007	2008	2009
▨ implicit electricity subsidies	2.30	2.10	1.80	1.50
☐ implicit natural gas subsidies	7.50	6.80	8.90	8.90
■ implicit oil subsidies	11.9	10.9	11.2	11.9

Source: IMF.
Note: Implicit subsidies (or hidden costs) are defined as the difference between actual receipts and the revenue that the energy company (for example, a utility involved in the distribution of electricity and natural gas) would receive were it to be in operation with cost-recovery tariffs based on efficient operation with normal losses and with full bill collection.

Fuel Prices and Road Sector Consumption in the Islamic Republic of Iran

Figure 15A.3   Domestic Retail Fuel Prices in the Islamic Republic of Iran, 2002–10

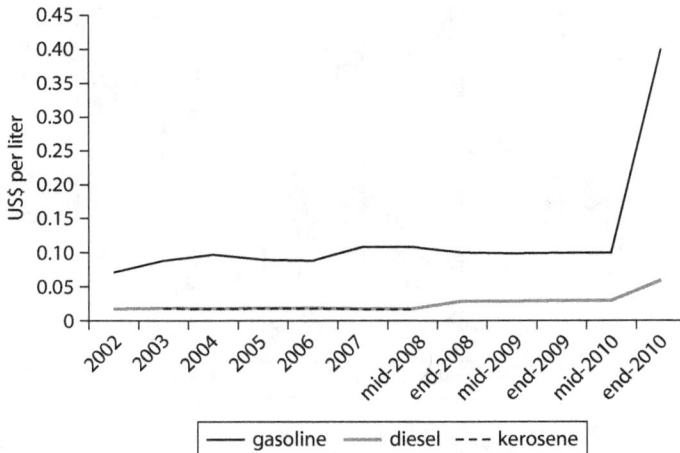

Sources: IMF and German Agency for International Cooperation (GIZ) reports, updated based on IMF 2011.

Figure 15A.4   Road Sector Diesel Consumption in the Islamic Republic of Iran, 1998–2008

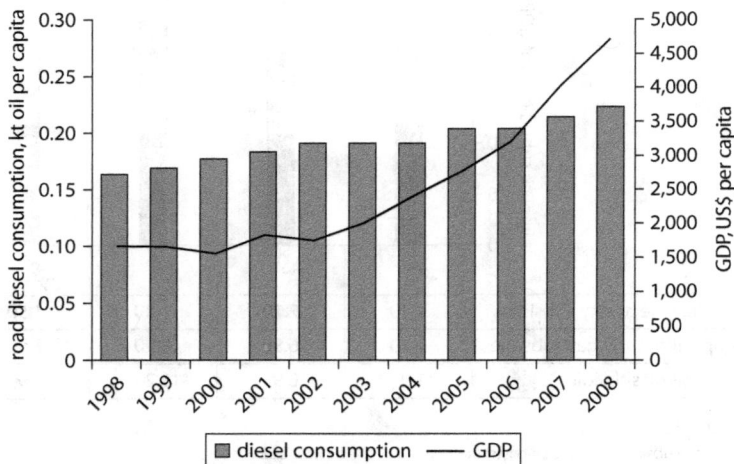

Source: World Bank, World Development Indicators.

Figure 15A.5   Road Sector Gasoline Consumption in the Islamic Republic of Iran, 1998–2008

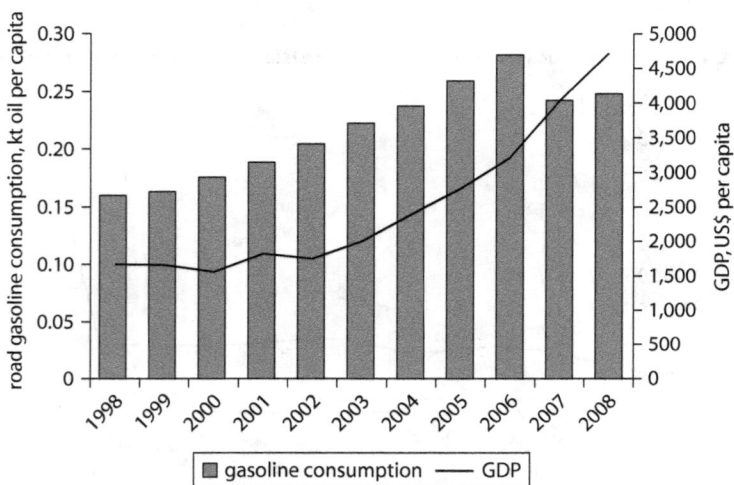

Source: World Bank, World Development Indicators.

Electricity Price and Power Consumption in the Islamic Republic of Iran

Figure 15A.6   Average Electricity Price in the Islamic Republic of Iran, 1998–2010

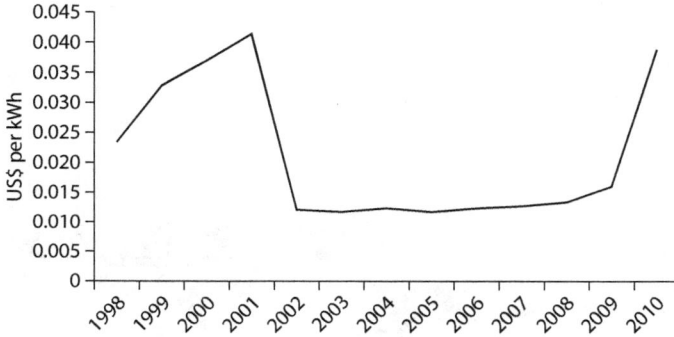

Source: Tavarin annual reports.
Note: Average residential and industrial price. kWh = kilowatt-hour.

Figure 15A.7   Power Consumption Per Capita in the Islamic Republic of Iran, 1998–2008

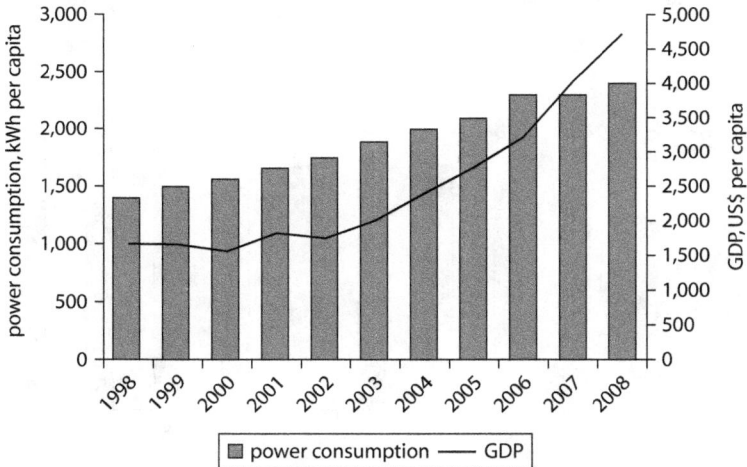

Source: World Bank, World Development Indicators.
Note: kWh = kilowatt-hour.

Poverty Impact Evidence from Household Surveys in the Islamic Republic of Iran

Figure 15A.8   Electricity Block Tariffs in the Islamic Republic of Iran, 2007 and 2010

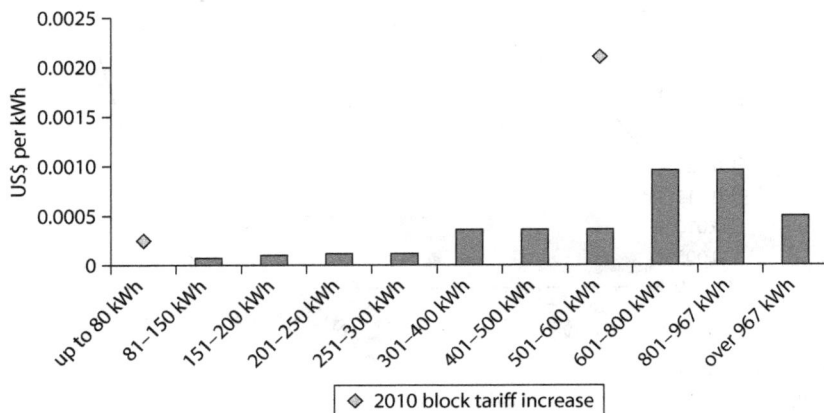

Source: World Bank 2007 based on 2007 tariff structure.
Note: kWh = kilowatt-hour.

Figure 15A.9   Benefit Incidence of Electricity Subsidies in the Islamic Republic of Iran, by Income Decile, 2005

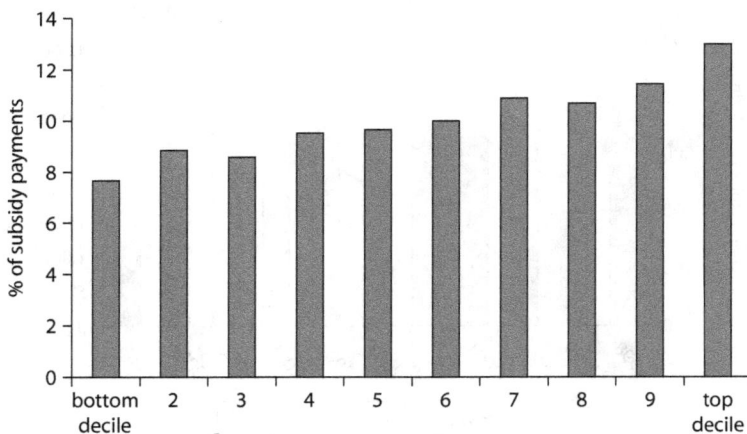

Source: World Bank 2007.

Figure 15A.10 Household Expenditure on Utilities in the Islamic Republic of Iran, by Income Decile, 2007

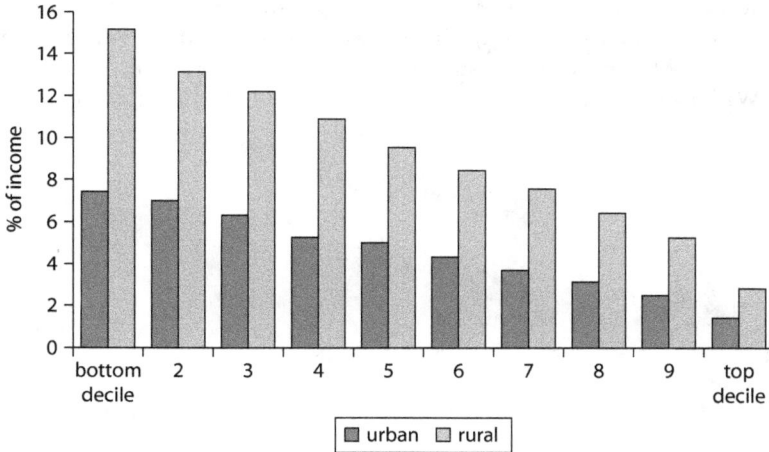

Source: Statistical Centre of the Islamic Republic of Iran, Vice Presidency for Planning and Strategic Supervision.
Note: Utility expenditures include expenditures on fuel, light, water, and sewerage, 2007.

Notes

1. See the Central Bank of the Islamic Republic of Iran website: http://www.cbi
.ir/default_en.aspx.
2. The Gini coefficient—the most commonly used measure of inequality of
income or consumption—varies between 0, which reflects complete equality,
and 1, which indicates complete inequality (World Bank, *World Development
Indicators*).

References

GIZ (German Agency for International Cooperation). n.d. International Fuel
Prices database. GIZ (formerly GTZ), Bonn. http://www.gtz.de/en/themen/
29957.htm.

Guillaume, Dominique, Roman Zytek, and Mohammad Reza Farzin. 2011. "Iran:
The Chronicles of the Subsidy Reform." Working Paper 11/167, IMF,
Washington, DC.

IMF (International Monetary Fund). 2009. "Islamic Republic of Iran: 2009 Article
IV Consultation." IMF, Washington, DC.

———. 2010. Retail domestic fuel prices data sheet, IMF, Washington, DC.
http://www.imf.org/external/pubs/ft/spn/2010/data/spn1005.csv.

————. 2011. "Islamic Republic of Iran: 2011 Article IV Consultation." Country Report 11/241, IMF, Washington, DC.

World Bank. 2007. "Iran Power Sector Reform." Power Sector Note, Sustainable Development Department, Middle East and North Africa Region, Energy Sector Management Assistance Program (ESMAP) and World Bank, Washington, DC.

CHAPTER 16

Nigeria

Incentives to Energy Subsidy Reforms

Nigeria is an oil exporting country with significant oil and gas resources. The oil sector has influenced significantly the growth contour of the country since 1970. In spite of cumulative efforts by successive governments, fuel subsidies remain an important socioeconomic issue. Explicit government fuel subsidy payments averaged US$450 million per year—averaging 0.4 percent of nominal non-oil gross domestic product (GDP) per year—from 2006 to 2009, according to International Monetary Fund (IMF) data. In the same period, implicit subsidies reached US$2.1 billion on average per year—or 1.9 percent of non-oil GDP (see figure 16A.1).

The Nigerian National Petroleum Corporation's refineries sell diesel at a subsidized price to product marketers even though the pump price of diesel is liberalized. These subsidies have been paid directly from the budget. Subsidies are provided at the refinery gate or at the point of product entry into the country. Fearing financial losses, independent fuel distributors stopped importing refined products in April 2009, causing significant economic dislocations. The combination of declining oil revenue and the high fiscal costs of subsidization prompted the government to launch renewed efforts to liberalize the fuel market (Nwafor, Ogujiuba, and Asogwa 2006; Adenikinju 2009; Kojima 2009; IMF 2011).

Although Nigeria is one of the largest oil producers in the world and has the seventh-largest gas reserves, the country has continued to suffer from a chronic shortage of power. This shortage has caused approximately 55 percent of the country's population to live without access to electricity (World Bank 2011a). Even when power is available, power outages continue to force over 90 percent of industrial and a significant number of residential consumers to install and run their own power generators at high cost to themselves, the Nigerian economy, and the environment. Years of poor maintenance in the power sector made matters worse (World Bank 2011a). In this regard, inadequate funding from non-cost-reflective tariffs and poor revenue collection rates played an important role. Tariff reforms—with the introduction of a multiyear tariff order in 2005—implied a significant reduction of hidden costs due to underpricing, from about 2.5 percent in 2005 to about 1.25 percent as of 2007 (see figure 16A.2)

The Nigerian economy has weathered both the global economic recession and its own domestic banking crisis reasonably well. The economy expanded more rapidly than expected in 2009 and continued to gain strength in 2010. The amnesty extended to rebels in the oil producing region led to a sharp recovery in oil production while non-oil GDP growth has remained high. Strong growth and targeted public expenditures have helped Nigeria make some progress toward achieving the United Nations' Millennium Development Goals (MDGs).[1] Nonetheless, policy slippages emerged during 2010: commitment to the oil revenue rule weakened, and foreign reserves fell even as oil prices have rebounded. The overall consolidated fiscal deficit contracted somewhat because of high oil revenues, but the non-oil primary deficit increased by 5 percentage points, to 32.2 percent of non-oil GDP. Most oil exporting countries are running fiscal surpluses. In that regard, the IMF supports proposed fiscal consolidation, including reductions in fuel subsidies (IMF 2011).

Reform Efforts

Fossil Fuels

In 2004, an oil revenue fiscal rule was adopted based on an informal political agreement among various levels of government but not rooted in legislation. The agreement provided for allocation of benchmark oil revenues, which are based on a budget benchmark oil price and projected oil production. The budget oil price is politically agreed on and approved by parliament. Any oil revenues in excess of the benchmark level are transferred into the "Excess Crude Account (ECA)" at the central bank in the names of the various tiers of government (IMF 2011).

In 2007, a Fiscal Responsibility Act was adopted—partly as an attempt to formalize the "voluntary" oil revenue-based fiscal rule. In September 2007, a political agreement was reached under which all states would pass fiscal responsibility legislation, but progress in promulgating similar legislation is limited so far (IMF 2011). In 2009, the IMF noted that the oil-price-based fiscal rule played an important role in Nigeria's success in weathering the financial crisis (IMF 2009). According to the IMF, the rule broke the link between public spending and oil prices and created an oil-savings cushion of US$18 billion, or 15 percent of non-oil GDP.

In January 2006, Nigeria established a Petroleum Support Fund to reimburse the difference between fuel import costs and revenues from selling fuels at subsidized prices. One objective was to create a level playing field whereby private sector companies could also import and participate in the sale of subsidized fuels. Although designed to be funded by the three tiers of government, the fund has in practice been financed since its inception by the federal government budget and the Domestic Excess Revenue Account.[2] From its inception through July 2008, the fund paid for subsidizing 33 billion liters of gasoline, kerosene, and diesel (this latter fuel type was covered by the fund in the first six months of 2006) at a total of US$7.2 billion. The Nigerian National Petroleum Corporation received about 80 percent of the total subsidy (PPPRA 2008; Kojima 2009). The sharp decline in oil prices starting in late 2008 tested the fiscal rule and the savings account for the first time. In addition, increased unrest in the oil-producing Niger Delta region led to a decline in oil production. Consequently, the authorities withdrew resources from the ECA to offset the shortfall in oil revenue. However, a recovery in oil prices and production during the second half of 2009 did not halt withdrawals from the ECA. The 80/20 rule—whereby 80 percent of the excess revenues saved in the ECA during the previous year would be disbursed regardless of movements in world oil prices—and other ad hoc withdrawals from the account have almost depleted the ECA and undermined its stabilization function (IMF 2011).

In 2008 and 2009, renewed efforts were made to liberalize petroleum and kerosene prices in order to free up significant resources and generate fiscal space (about 2 percent of non-oil GDP) for higher-priority spending (IMF 2009; Kojima 2009).

Fuel subsidies have resulted in substantial loss of revenue and growth in road fuel consumption (see figure 16A.4) because low prices do not signal the real cost of consumption; subsidies contributed to the collapse of local refineries because the price of fuel did not reflect the cost of supply.

Electricity

- Comprehensive power sector reform was launched in 2005. The federal government promulgated the Electricity Power Reform Act, which sought to restructure the power sector and open it to private sector investment. It established a framework for sector reforms to establish a market-oriented industry structure and required opening access to the grid on a nondiscriminatory basis and to spur competition among producers. A multiyear tariff order was introduced in 2005. In addition, the National Electricity Power Authority, formed in 1972, was broken up and split into 18 companies[3] (forming a new holding company, the Power Holding Company of Nigeria, PHCN). The PHCN has embarked on a number of interventions since 2007, including a national awareness campaign. The Nigerian authorities also established the Nigeria Electricity Regulatory Commission as an independent regulatory agency to ensure efficient and equitable growth of the electric power sector (Akinlo 2009; World Bank 2011a).

- From 2007 to 2009, progress on power sector reform was stalled as the administration declared a national emergency in the power sector to address the many short-term challenges facing the nation's ailing power supply. The reform agenda was relaunched in August 2010 (World Bank 2011a).

- For the future, the authorities are proposing a sovereign wealth fund, the Nigerian Sovereign Investment Authority (IMF 2011). It will have three separate components: a stabilization fund, a fund for future generations, and a domestic infrastructure fund. The proposed legislation would establish a governing council that would include representatives from civil society organizations, academics, and other private sector representatives. A minimum of 20 percent of surplus oil revenues would be allocated to each of the three components in any given year. The infrastructure fund would finance investments in power generation, distribution and transmission, water and sewage treatment and delivery, roads, port, rail and airport facilities, and other infrastructure-related projects within Nigeria (IMF 2011).

Poverty Alleviation Measures

- For the bottom income quintile, of which only 10 percent of households have access to electricity, increasing block tariffs do not offer any protection (see figures 16A.8 and 16A.9). At the same time, it is the bottom quintile that spends the highest share of its total monthly expenditures on electricity (see figure 16A.10).

- The National Poverty Eradication Program (NAPEP), as approved by the Federal Executive Council in 2001, has the major mandate of multisectorally monitoring and coordinating all poverty eradication efforts in Nigeria with a view to harmonizing these efforts and bringing about the focus and complementation required at all levels to ensure better delivery, maximum impact, effective utilization of resources, and easy review (Bindir 2001).

- A Conditional Grants Scheme (CGS) was introduced in 2006 to make progress in achieving the MDGs. The CGS has been designed to address the MDGs at the local level by providing financial and technical support to scale up the MDG-related activities of state and local governments. Conditional grants channel funds, technical assistance, and best practices from the federal government to subnational governments. Their goal is to reduce the fiscal constraints that states and local governments face while improving their capacity and demand for effective service delivery (Government of Nigeria 2010).

- NAPEP's pilot Conditional Grant and Cash Transfer (CCT) program was established in 2007. The World Bank provided technical support, implementation started in early 2008 in 12 states, and it is active to date. The benefit structure consists of a cash transfer based on the number of children per household. The monthly payments are done through microfinance agencies and local community banks (World Bank 2011b).

- In 2011, the IMF noted that the reallocation of resources to capital projects would be more supportive of long-term growth and poverty reduction. In that regard, an Expenditure Review Committee, with participation from government, academia, civil society, and the private sector, has been established to suggest long-term savings on recurrent outlays to support fiscal consolidation and release resources for capital projects. The authorities recognize that insufficient implementation capacity—not the availability of financing—has so far been the major constraint to increasing spending on infrastructure, and they are working to increase the capacity to implement capital projects (IMF 2011).

Key Lessons Learned

Subsidies have had a particularly harmful effect on the competitiveness of existing refineries and have discouraged entry of new firms into the industry. Nigeria's downstream sector has become increasingly inefficient and undercapitalized.

The social safety system appears fragmented and in need of improvement, especially in the area of CCTs. The undertaking of a government-requested Poverty and Social Impact Analysis is an important step toward understanding the impact on society of the subsidy phaseout and learning how to improve the safety nets to provide effective remedial and long-term measures.

Despite being the world's eighth-biggest crude oil exporter, Nigeria still imports around 85 percent of its petroleum products because of poorly managed refineries and insufficient refining capacity. Domestic production is far from meeting demand, with imports making up for the gap. Moreover, fuel shortages have opened up opportunities for selling fuels on the black market.

Annex 16.1 Nigeria Case Study Figures

INCOME LEVEL: Low-income
REGION: Sub-Saharan Africa
ENERGY NET IMPORTER/EXPORTER: Net exporter
SUBSIDIES: Gasoline, kerosene
PHASING OUT SUBSIDIES: Ongoing

Fiscal Burden of Energy Subsidy in Nigeria

Figure 16A.1   Explicit and Implicit Fuel Subsidies in Nigeria, 2006–10

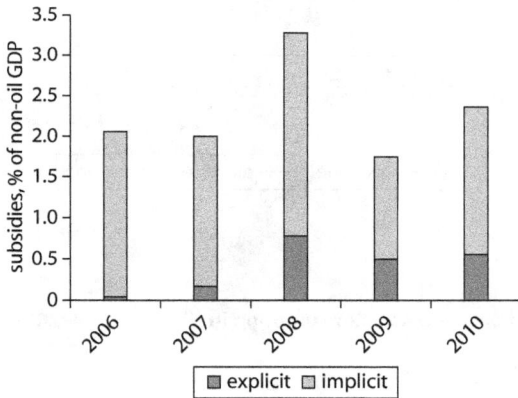

Source: IMF 2011.

Figure 16A.2   Implicit Power Sector Subsidies in Nigeria, 2005–09

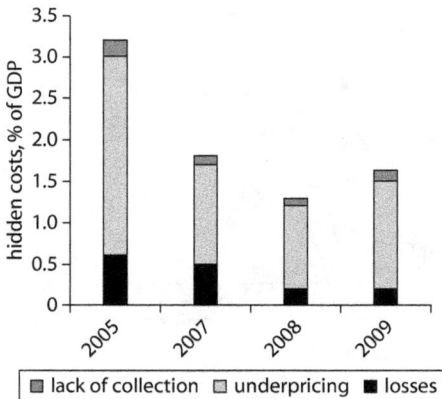

Source: AICD 2010.
Note: Implicit subsidies (or hidden costs) are defined as the difference between actual receipts and the revenue that the energy company (for example, a utility involved in the distribution of electricity and natural gas) would receive were it to be in operation with cost-recovery tariffs based on efficient operation with normal losses and with full bill collection.

Fuel Prices and Road Sector Consumption in Nigeria

Figure 16A.3   Domestic Retail Fuel Prices in Nigeria, 2002–10

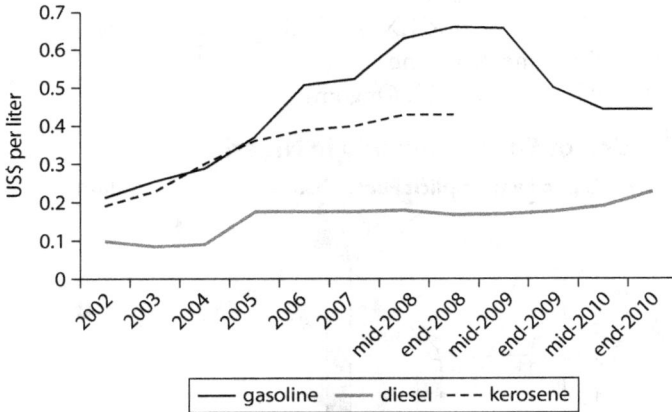

Source: IMF 2010; GIZ n.d.; PPPRA 2008.

Figure 16A.4   Road Sector Diesel Consumption in Nigeria, 1998–2008

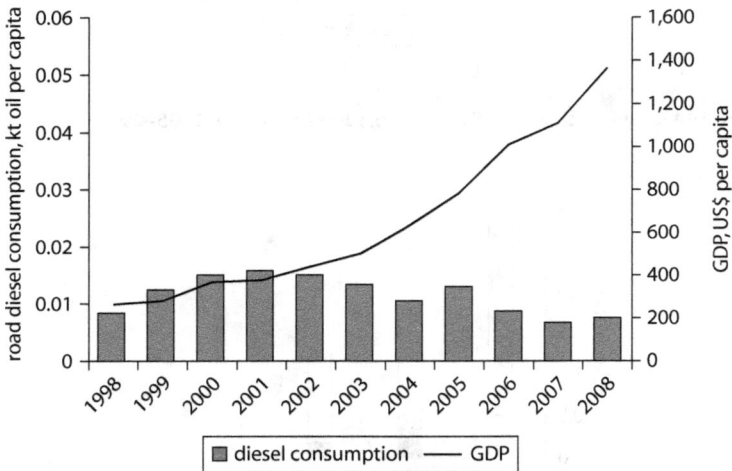

Source: World Bank, World Development Indicators.

Figure 16A.5   Road Sector Gasoline Consumption in Nigeria, 1998–2008

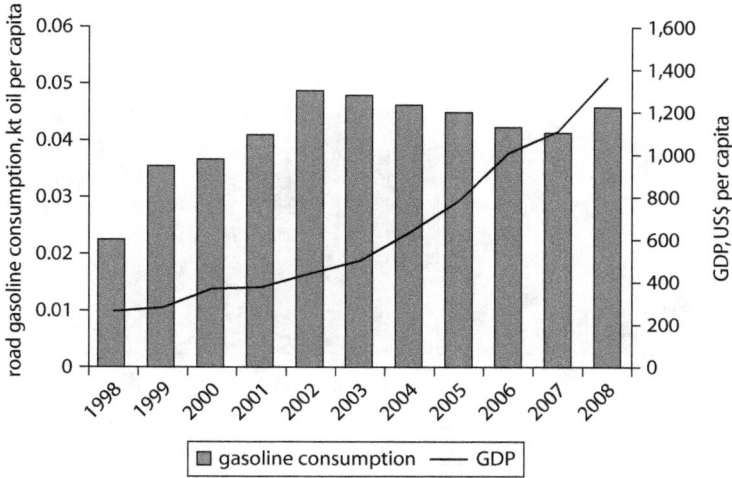

Source: World Bank, World Development Indicators.

Electricity Price and Power Consumption in Nigeria

Figure 16A.6   Average Electricity Price in Nigeria, 2002–10

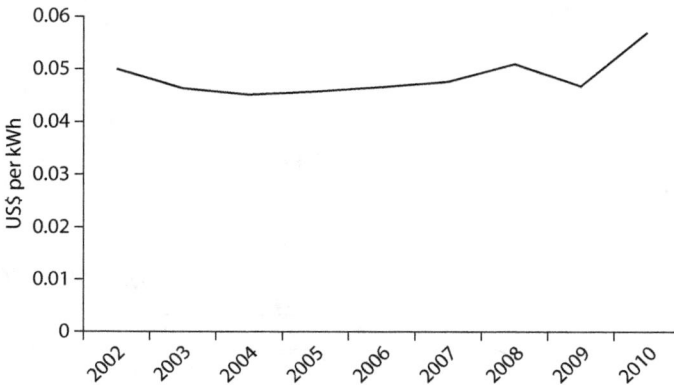

Source: World Bank, World Development Indicators.
Note: kWh = kilowatt-hour.

Figure 16A.7   Power Consumption Per Capita in Nigeria, 1998–2008

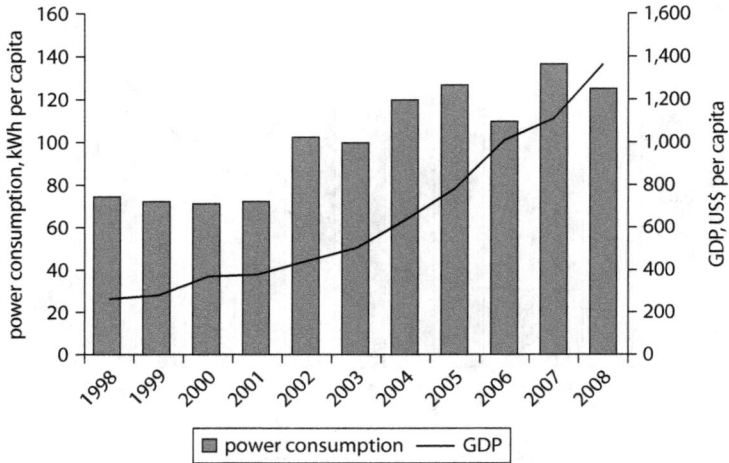

Source: World Bank, World Development Indicators.
Note: kWh = kilowatt-hour.

Poverty Impact Evidence from Household Surveys in Nigeria

Figure 16A.8   Electricity Block Tariffs in Nigeria, 2011

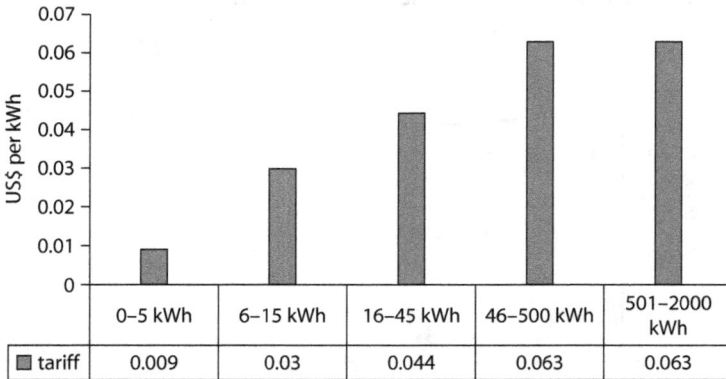

	0–5 kWh	6–15 kWh	16–45 kWh	46–500 kWh	501–2000 kWh
tariff	0.009	0.03	0.044	0.063	0.063

Source: AICD database.
Note: kWh = kilowatt-hour.

Figure 16A.9   Access to Electricity in Nigeria, by Income Quintile, 2011

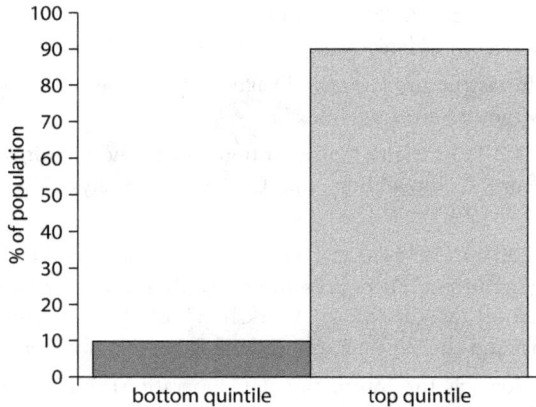

Source: AICD database.

Figure 16A.10   Electricity Expenditure in Nigeria, by Income Quintile, 2011

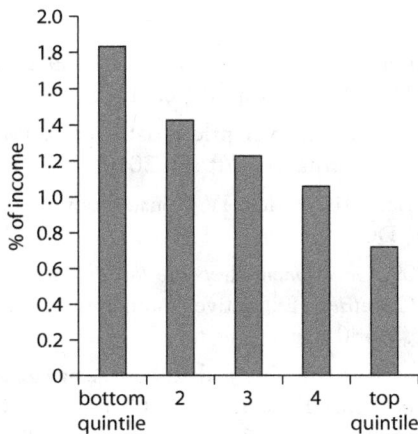

Source: AICD database.

Notes

1. For more information about the Millennium Development Goals, see the website at http://www.un.org/millenniumgoals/.

2. The Domestic Excess Revenue Account is a sovereign wealth fund established in 2004 to accrue revenue derived from crude oil sales, petroleum profit taxes, and royalties over and above the budgeted benchmark.

3. Six generation companies, the Transmission Company of Nigeria, and 11 distribution companies.

References

Adenikinju, Adeola. 2009. "Energy Pricing and Subsidy Reforms in Nigeria." http://www.oecd.org/dataoecd/58/61/42987402.pdf. Accessed July 2011.

AICD (Africa Infrastructure Country Diagnostic) database. Various years. http://www.infrastructureafrica.org.

Akinlo, A. E. 2009. "Electricity Consumption and Economic Growth in Nigeria: Evidence from Cointegration and Co-Feature Analysis." *Journal of Policy Modeling* 31 (5): 681–93.

Bindir, Umar. 2001. "The National Poverty Eradication Programme (NAPEP) Monitoring Strategies." Paper presented at the "Community-Based Monitoring and Evaluation Methodology Workshop" of Food Basket Foundation International and the World Bank, Federal Republic of Nigeria, June 10–14.

GIZ (German Agency for International Cooperation). n.d. International Fuel Prices database. GIZ (formerly GTZ), Bonn. http://www.gtz.de/en/themen/29957.htm.

Government of Nigeria. 2010. "Partnering to Achieve the MDGs: The Story of Nigeria's Conditional Grants Scheme 2007–2010." Government of Nigeria, Abuja.

IMF (International Monetary Fund). 2009. "Nigeria: 2009 Article IV Consultation." Country Report 09/315, IMF, Washington, DC.

———. 2010. Retail domestic fuel prices data sheet, IMF, Washington, DC. http://www.imf.org/external/pubs/ft/spn/2010/data/spn1005.csv.

———. 2011. "Nigeria: 2010 Article IV Consultation." Country Report 11/57, IMF, Washington, DC.

Kojima, Masami. 2009. *Government Response to Oil Price Volatility: Experience of 49 Developing Countries*. Extractive Industries for Development Series. Washington, DC: World Bank.

Nwafor, Manson, Kanayo Ogujiuba, and Robert Asogwa. 2006. "Does Subsidy Removal Hurt the Poor? Evidence from Computable General Equilibrium Analysis." Research Paper 2, African Institute for Applied Economics, Enugu, Nigeria.

PPPRA (Petroleum Products Pricing Regulatory Agency). 2008. "Report on the Administration of the Petroleum Support Fund: January 2006–July 2008." http://www.pppra-nigeria.org. Accessed July 2011.

World Bank. 2011a. "Nigeria: Power Sector Guarantees Project." Project Information Document, Report AB6379, World Bank, Washington, DC.

———. 2011b. CCT Program Profile–Nigeria." http://web.worldbank.org/WBSITE/EXTERNAL/TOPICS/EXTSOCIALPROTECTION/EXTSAFETYNETSANDTRANSFERS/0,,contentMDK:22063621~pagePK:148956~piPK:216618~theSitePK:282761~isCURL:Y,00.html. Accessed July 2011.

CHAPTER 17

Republic of Yemen

Incentives to Energy Subsidy Reforms

The Republic of Yemen's oil reserves, on which the economy depends heavily, are expected to run out within a decade in the absence of new oil discoveries. With the collapse of international oil prices in late 2008, the country's oil revenues—accounting for around 60 percent of government revenue and over 90 percent of export revenue—have declined (IMF 2011; World Bank 2010, 2011a).

Petroleum subsidies, on average, have amounted to 8–10 percent of gross domestic product (GDP) (see figure 17A.1)—accounting for more than total spending on education, health, and social transfers combined (IMF 2009; World Bank 2011b). The subsidy magnitude affects both households and economic sectors. Households consume only 10 percent of all fuel products; the rest are consumed as intermediate inputs in agriculture, industry, and services. The largest share of fuel subsidies goes to diesel, which made up more than two-thirds (69 percent) of all subsidized fuels in 2009. In addition, 14 percent of fuel subsidies were gasoline subsidies, and the remainder was split between liquefied petroleum gas (LPG), kerosene, and jet fuel (World Bank 2011b). The total amount of energy subsidies in 2008 increased to more than 14 percent of GDP because of a sharp increase in the international oil price (with domestic prices staying constant) and the growth in domestic energy demand (see figures 17A.1 and 17A.2). The overall fuel subsidy bill depends on changes in the international prices of petroleum products because the domestic price is fixed. Therefore, the record international fuel prices for the first half of 2008 raised the fuel subsidy bill (IMF 2009). A large supplementary budget was approved toward the end of 2008 to validate additional spending on fuel subsidies and an increase in wages, pensions, and social welfare transfers.

The subsidy picture in the electricity sector is quite complicated. There is an apparent or explicit subsidy of YRl 5 per kilowatt-hour (kWh), which is based on the difference between the total nonfuel operating expenses of YRl 17 per kWh and the average selling price of YRl 12 per kWh. However, the electricity sector receives full subsidy for its investment costs (estimated at YRl 6 per kWh) and for more than 80 percent of its fuel cost (estimated at YRl 13 per kWh). Therefore, the total subsidy is at least YRl 24 per kWh (about US$0.12 per kWh). This means that presently the electricity price covers only one-third of the cost while the other two-thirds are subsidized (World Bank 2008). The electricity subsidy represented 0.1 percent of GDP in 2006–10 (see figure 17A.1).

Reform Efforts

An attempt to increase prices for diesel, gasoline, and kerosene by 144 percent in July 2005 resulted in riots against the policy that raised prices, leaving 39 people dead and hundreds injured. Since then, no reform was attempted until more recently. As oil prices began to decline in the latter part of 2008, measures were taken to compress expenditures, including fuel subsidies, by raising the diesel price to industrial users.

The Yemeni government took steps to initiate comprehensive petroleum subsidy reform by increasing the price of fuel by about 25 percent in 2010, while subsidies for oil derivatives were reduced. The latest adjustments (October 2010) increased diesel fuel prices by about 11 percent and regular gasoline prices by about 7 percent, as can be seen in figure 17A.2 (World Bank 2011b). However, the continuous rise of international prices for oil products reduced the savings obtained from the subsidy reductions: the 2010 savings were expected to be 0.8 percent of GDP (World Bank 2011b).

In 2011, the International Monetary Fund (IMF) noted that the authorities had been in the process of increasing fuel prices with the aim of an eventual elimination of subsidies. At the same time, the authorities were increasing social transfers to protect the most vulnerable segments of the population to counteract the rise in fuel prices (see figure 17A.2), the IMF (2011) noted.

Poverty Alleviation Measures

Evidence from Household Surveys

Regarding the structure of the electricity block tariffs, the first block of 200 kWh per month in urban areas is found to be "too high" (see figure 17A.7). According to a World Bank assessment, a better-targeted lifeline rate would benefit the poor more (World Bank 2005). The immediate impact of reduced subsidies on the poorest groups is likely to be that they are pushed back to using biomass, which not only has undesirable impacts on the environment (pressure on woodland resources) and on health (respiratory diseases), but will also increase time burdens on poor families for fuel collection (World Bank 2005).

The pattern of energy use is relatively similar in rural and urban areas, with the notable exception of electricity, to which only a low percentage of poor households have access in rural areas (less than 10 percent for the lowest two deciles, compared with 50 percent of the top deciles) (see figure 17A.8). Accordingly, the most burdensome sources of energy in rural areas are kerosene, biomass, and LPG, all of which display a progressive pattern, decreasing as income increases. In the case of urban areas, electricity is instead the most burdensome source of energy in household budgets (see figure 17A.9).

More than 77 percent of the direct subsidies on petroleum products accrue to the nonpoor, while only 23 percent goes to the poor. In fact, households in the highest income decile receive 40 percent of the diesel subsidy (see figure 17A.10). Households in the two lowest income deciles receive only 2 percent of the total diesel subsidy. Since kerosene is mainly consumed by poor households, the amount of the total subsidy for kerosene is distributed more equally. The World Bank report points out that the petroleum product subsidies result in a real cost to society (that is, deadweight loss) because the incremental cost of the subsidy to government exceeds the incremental benefit to consumers (World Bank 2005). In 2009, the IMF noted that the fuel subsidy system should be phased out in combination with an increase in, and better targeting of, social welfare transfers (IMF 2009).

If prices of all petroleum products were raised to their economic levels, then the estimate of the impact on the different income deciles is as

shown in figure 17A.11.[1] Low-income households would bear a relatively higher burden than higher-income households if subsidies were to
be removed. This is particularly true for kerosene because it is used
mostly by the poor. The removal of subsidies for diesel fuel, however,
would be more costly to high-income households. Again, this is due to the
fact that richer deciles tend to consume more diesel than poorer deciles
(see also figures 17A.8 and 17A.9). The aggregate effects of eliminating
subsidies on most of the fuels are regressive—that is, the poor's energy
expenditures would rise by a greater proportion than the nonpoor's
(World Bank 2005).

The model results show that reducing fuel subsidies without taking
additional measures will increase poverty for both rural and urban households (World Bank 2005). The use of all savings from subsidy removal for
direct transfers can greatly reduce the negative impact on households, but
it will not provide necessary growth impulses for sustainable development. Accordingly, a combination of fiscal deficit reduction, social transfers, and investments is the most promising reform strategy.

The model considers two major scenarios: an accelerated reform (subsidy removal within one year) and a gradual reform (phasing out of subsidies over three years). Under the first scenario, the fiscal deficit was
expected to be reduced from 6.9 percent of GDP in 2010 to 3.5 percent
in 2011 and the surplus from the reforms was estimated to be YRl 215
billion. Under the gradual-reform scenario, the fiscal deficit was expected
to be reduced from 5.8 percent of GDP in 2011 to 4.6 percent of GDP
in 2012 and to 3.5 percent in 2013; surplus from the gradual reforms was
estimated to be YRl 72 billion per year for 2011–13. As for the compensation to the poor, the results of the model showed that under the
accelerated-reform scenario, the total annual cost of compensating the
poorest 30 percent would be about YRl 17.6 billion per year for 2010–15;
for the gradual reforms, the amount required for compensation would be
YRl 12.3 billion per year (World Bank 2005).

Social Safety Nets

The Republic of Yemen is one of the poorest countries in the Middle East
and North Africa region, and it faces challenges including rapid population growth, limited institutional capacity and outreach of the state,
rapidly depleting water reserves, poor infrastructure, limited human
development, and acute inequality issues. The country has been hit hard
by the global food and financial crises. The rise in food prices in 2007 and
2008 had serious consequences for household budgets and hence for the

desired reduction in the poverty rate. Although considered inefficient and distortive, the untargeted energy subsidy scheme constitutes the largest public safety net and remains critical for the poor. Removing energy subsidies without effective alternatives in place would increase poverty rates substantially (World Bank 2005, 2010).

The Republic of Yemen has in place successful community-driven programs under the Social Fund for Development (SFD) and the Public Works Project (PWP) as well as the provision of defined benefits through cash transfers from the Social Welfare Fund (SWF). Altogether, public financing for these programs is still modest, representing only 0.6 percent to 1 percent of GDP (until 2008).[2] The SFD and PWP activities provide medium- to long-term benefits to poor communities by supporting access to social services and economic opportunities, but the SWF program is the only national program mandated to reach chronically poor households with immediate safety net support through cash assistance. In the past, the SWF has suffered from a combination of low benefits, poor targeting, and inefficient administrative and operational processes, resulting in chronically low coverage of the poorest and little impact on overall poverty (World Bank 2010).[3]

In response to the food price crisis, the government initiated an emergency workfare program funded through a grant from the Global Food Crisis Response Program. This program is implemented by the SFD and was expected to reach poor households within the communities most seriously affected by the food crisis. A second action—in response to the food price crisis and the multiple shocks Yemen has recently experienced—has been to expand the coverage and increase benefits from the SWF (World Bank 2010).

Other social programs include government-sponsored institutions such as the Disability Fund and the Fund for Productive Families—which are, however, negligible in terms of actual expenditure levels. As in other Islamic countries, the Republic of Yemen has an informal safety net in the form of Zakāt (alms giving). At present, Zakāt collection officially amounts to 0.2 percent of GDP (World Bank 2010).

Key Lessons Learned

Social safety nets in the Republic of Yemen need to further develop before the country phases out energy subsidies. Compensating the poorest of the poor for their losses during reform will be important for success (World Bank 2010, 2011b).

An important lesson is the 2005 unrest from attempts to phase out subsidies, which not only were unsuccessful but also left many protesters dead. Formulating a clear reform agenda involving all stakeholders in the process is under way. New and more effective outreach methods for informing the public about reforms and compensation packages are also being proposed.

Annex 17.1 Republic of Yemen Case Study Figures

INCOME LEVEL: Low-income
REGION: Middle East and North Africa
ENERGY NET IMPORTER/EXPORTER: Net exporter
SUBSIDIES: Diesel, gasoline, electricity
PHASING OUT SUBSIDIES: Ongoing

Fiscal Burden of Energy Subsidy in the Republic of Yemen

Figure 17A.1   Explicit Budgetary Energy Subsidies in the Republic of Yemen, 2000–10

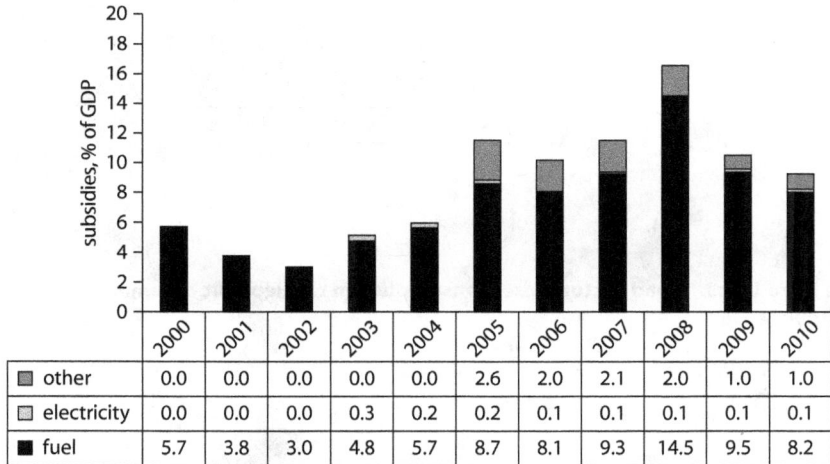

	2000	2001	2002	2003	2004	2005	2006	2007	2008	2009	2010
other	0.0	0.0	0.0	0.0	0.0	2.6	2.0	2.1	2.0	1.0	1.0
electricity	0.0	0.0	0.0	0.3	0.2	0.2	0.1	0.1	0.1	0.1	0.1
fuel	5.7	3.8	3.0	4.8	5.7	8.7	8.1	9.3	14.5	9.5	8.2

Source: IMF reports, various years.

Fuel Prices and Road Sector Consumption in the Republic of Yemen

Figure 17A.2   Domestic Retail Fuel Prices in the Republic of Yemen, 2002–10

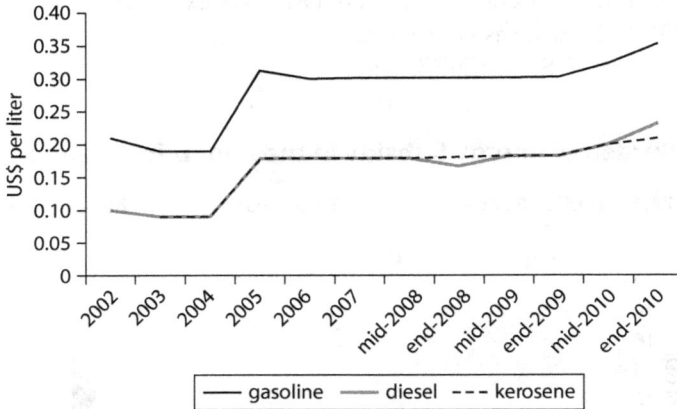

Source: IMF reports and GIZ n.d., updated.

Figure 17A.3   Road Sector Diesel Consumption in the Republic of Yemen, 1998–2008

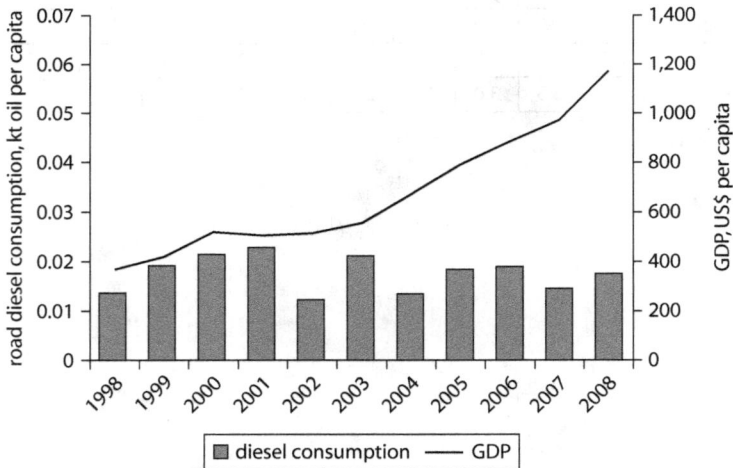

Source: World Bank, World Development Indicators.

Figure 17A.4   Road Sector Gasoline Consumption in the Republic of Yemen, 1998–2008

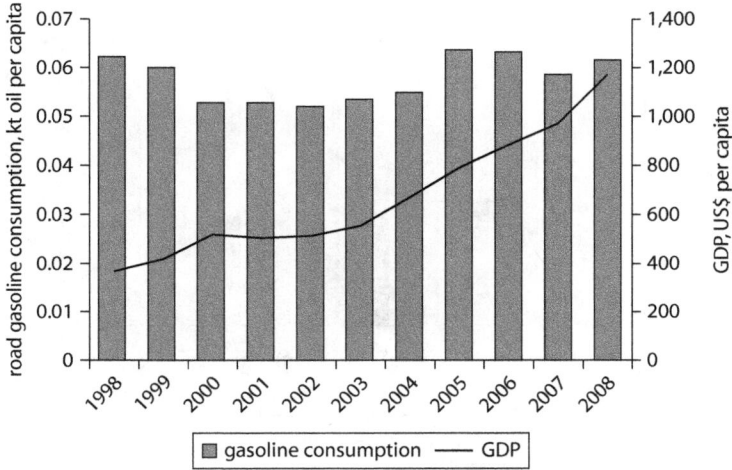

Source: World Bank, World Development Indicators.

Electricity Price and Power Consumption in the Republic of Yemen

Figure 17A.5   Average Electricity Price in the Republic of Yemen, 1999–2009

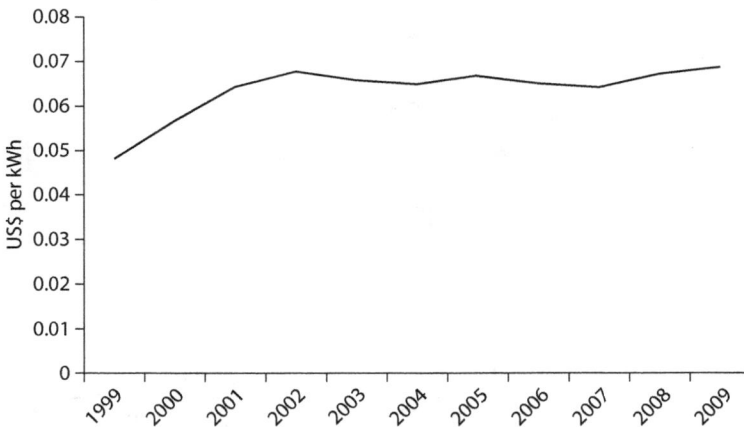

Source: PEC 2011.
Note: kWh = kilowatt-hour.

Figure 17A.6   Power Consumption Per Capita in the Republic of Yemen, 1998–2008

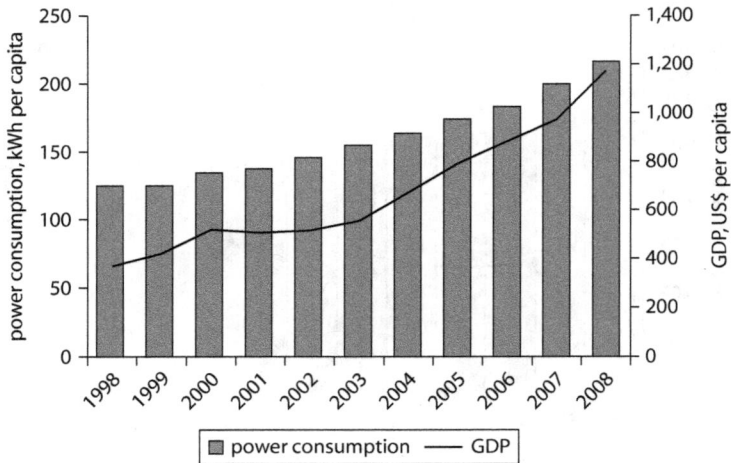

Source: World Bank, World Development Indicators.
Note: kWh = kilowatt-hour.

Poverty Impact Evidence from Household Surveys in the Republic of Yemen

Figure 17A.7   Electricity Block Tariffs in the Republic of Yemen, 2010

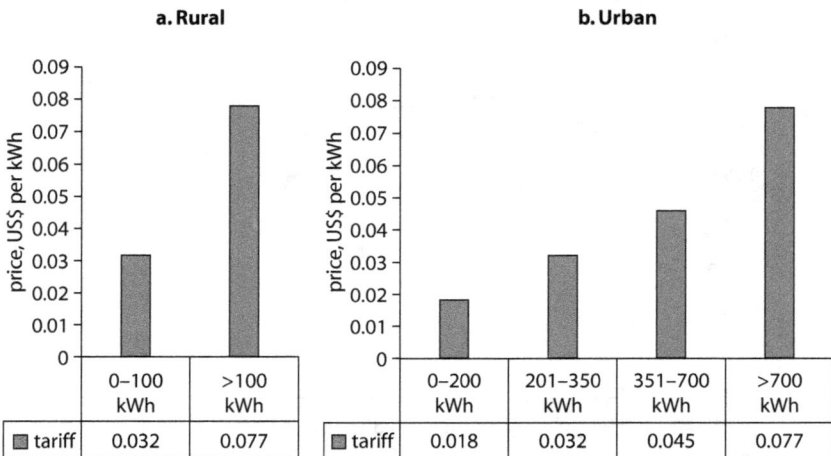

a. Rural

	0–100 kWh	>100 kWh
tariff	0.032	0.077

b. Urban

	0–200 kWh	201–350 kWh	351–700 kWh	>700 kWh
tariff	0.018	0.032	0.045	0.077

Source: PEC 2011.
Note: kWh = kilowatt-hour.

Figure 17A.8   Household Energy Use in the Republic of Yemen, by Income Decile, 2003

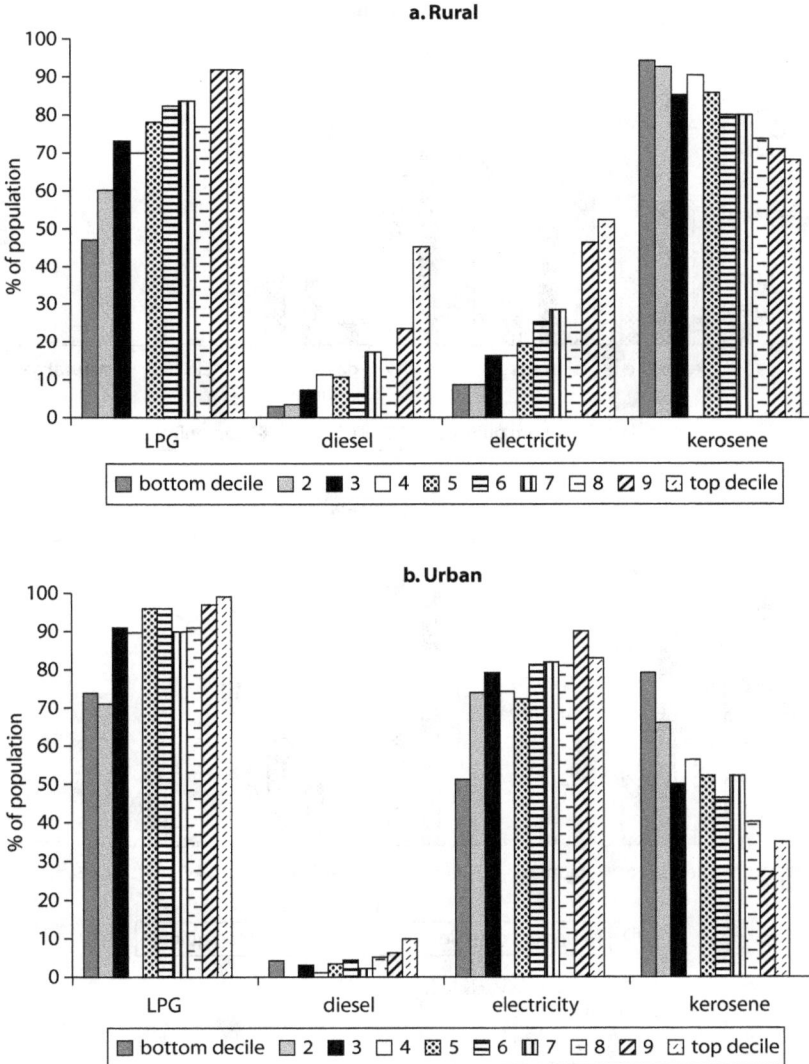

a. Rural

b. Urban

Source: World Bank 2005, based on 2003 household energy survey.
Note: LPG = liquefied petroleum gas.

Figure 17A.9   Household Energy Expenditure in the Republic of Yemen, by Income Quintile, 2003

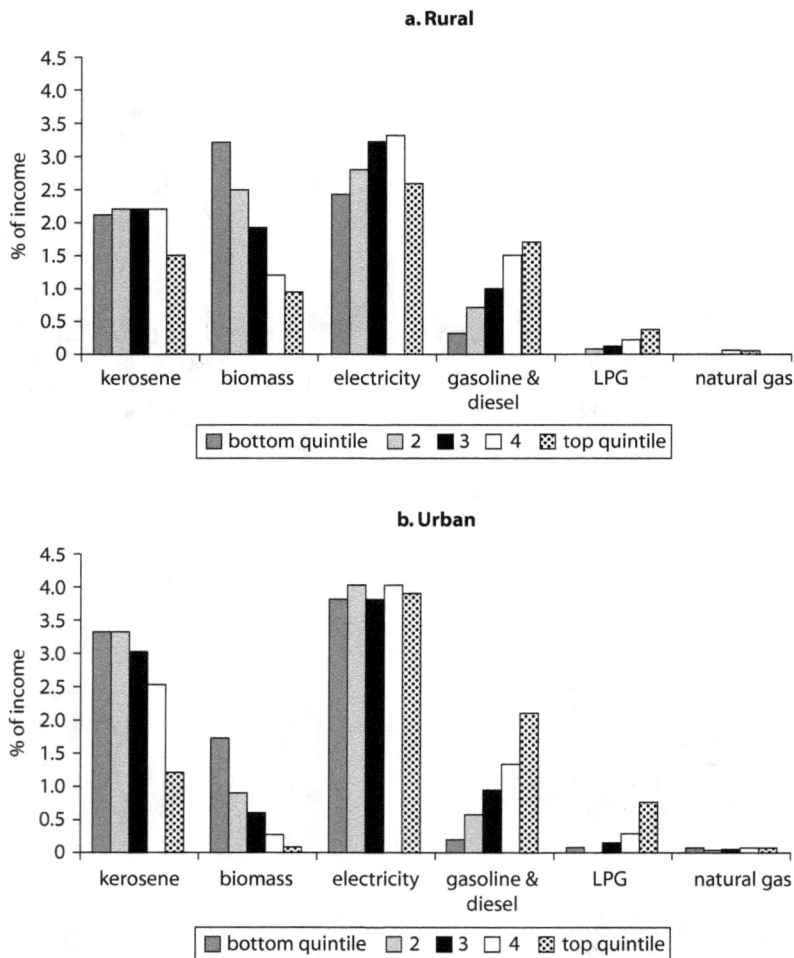

a. Rural

b. Urban

Source: World Bank 2005, based on 2003 household energy survey.
Note: LPG = liquefied petroleum gas.

Figure 17A.10   Benefit Incidence of Energy Subsidies in the Republic of Yemen, by Income Decile, 2003

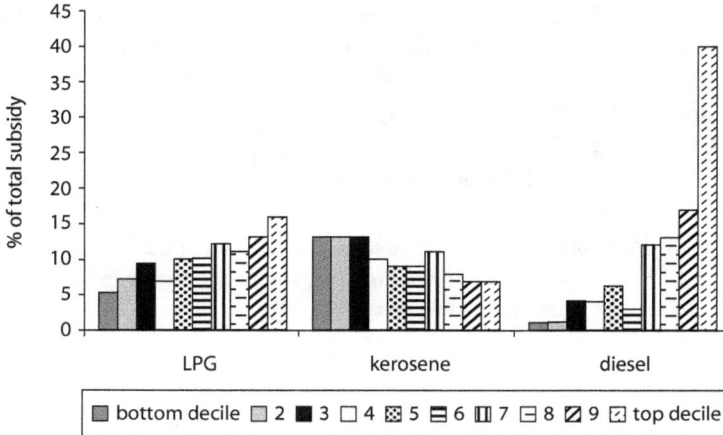

Source: World Bank 2005, based on 2003 household energy survey.
Note: LPG = liquefied petroleum gas.

Figure 17A.11   Welfare Impact of Fossil Fuel Subsidy Removal in the Republic of Yemen, by Income Decile, 2003

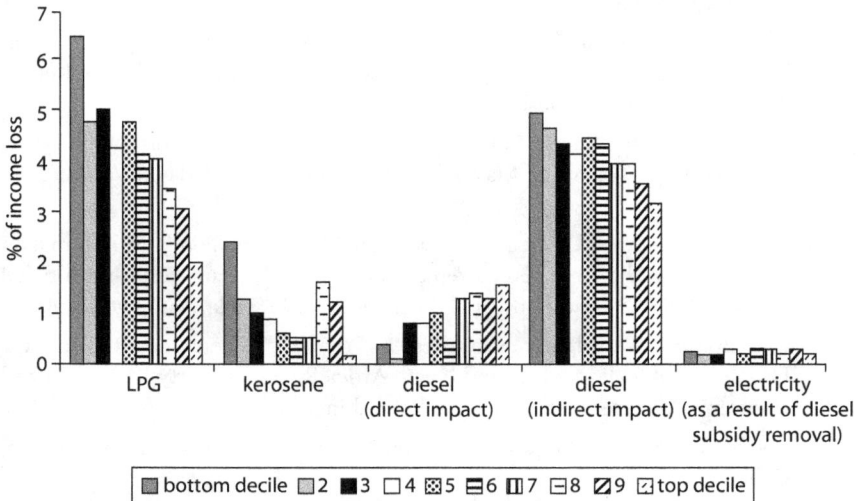

Source: World Bank 2005, based on 2003 household energy survey.
Note: LPG = liquefied petroleum gas.

Notes

1. The *direct* impact of adjusting fuel prices on the various income groups can be calculated by applying the difference between the economic price and the actual price paid for each of the various fuels to the consumption data for each income decile. Since almost all kerosene and LPG is purchased directly by households, the *indirect* impacts attributable to these fuels are minimal (World Bank 2005).

2. While the SFD and PWP are highly dependent on external funding (donors), the SWF is entirely funded by the public budget.

3. The World Bank's 2007 Poverty Assessment concluded that the SWF covered only 13 percent of the poorest population, and nearly two-thirds of beneficiaries were above the national poverty line.

References

GIZ (German Agency for International Cooperation). n.d. International Fuel Prices database. GIZ (formerly GTZ), Bonn. http://www.gtz.de/en/themen/29957.htm.

IMF (International Monetary Fund). 2009. "Republic of Yemen: 2008 Article IV Consultation." Country Report 09/100, IMF, Washington, DC.

———. 2011. "Republic of Yemen: Program Note." IMF, Washington, DC.

PEC (Public Electricity Corporation Yemen) database. 2011. http://www.pec.com.ye. Accessed June–July 2011.

World Bank. 2005. "Household Energy Supply and Use in Yemen: Volume I, Main Report." Energy Sector Management Assistance Programme (ESMAP) Report 315/05, World Bank, Washington, DC.

———. 2008. "Yemen Energy Subsidy Reform." Unpublished report by the World Bank Visiting Mission of May 24–June 1, 2008. World Bank, Washington, DC.

———. 2010. "Proposed Grant in the Amount of SDR 6.6 million (US$10.0 million equivalent) to the Republic of Yemen for a Social Welfare Fund Institutional Support Project." Project appraisal document, World Bank, Washington, DC.

———. 2011a. "Petroleum Subsidies in Yemen: Leveraging Reform for Development." Middle East and North Africa Region, Poverty Reduction and Economic Management, World Bank, Washington, DC.

———. 2011b. "Yemen Quarterly Economic Review." World Bank, Sana'a, Yemen, Office.

PART 4

Group D Countries:
Net Energy Exporter and High Income

Macroeconomic and Social Challenges

- Whereas Malaysia and Mexico are characterized by increasing levels of income per capita, Argentina has only slightly increased its gross domestic product (GDP) per capita. Income inequality has been increasing over time in both Malaysia and Mexico, whereas for Argentina it has declined over time. Mexico displays one of the highest GDPs per capita but also the highest level of income inequality (as reflected by the Gini index) in Group D (see figures P4.1 and P4.2).

- Argentina and Mexico are characterized by decreasing fiscal deficits. However, whereas Mexico managed to reduce public debt over time, Argentina has substantially increased it (see figures P4.3 and P4.4). Malaysia has worsened in terms of both fiscal budget management (moving from a small surplus to a deficit) and public debt, which has increased over time.

Fossil Fuel Dependence

- All countries in Group D have either increased or kept constant the percentage of electricity generated from fossil fuels, but they have significantly decreased their net exports of energy over time (see figures P4.5 and P4.6).

- Malaysia relies almost entirely on electricity generated from fossil fuel production and has reduced its net export of energy over time, from 80 percent of its own use in 1998 to about 30 percent in 2008. Mexico has been characterized by a similar pattern of dependence on fossil fuels, keeping a high percentage (around 80 percent) of energy production from fossil fuels and reducing energy net export by almost half— from about 60 percent to less than 30 percent. Argentina is characterized by the highest increase in the percentage of fossil fuels used to generate electricity as well as the highest decrease in net export, dropping from 33 percent in 1998 to less than 10 percent in 2008.

Income and Inequality Trends for Group D

Figure P4.1   GDP Per Capita, Group D Countries, 1998–2008

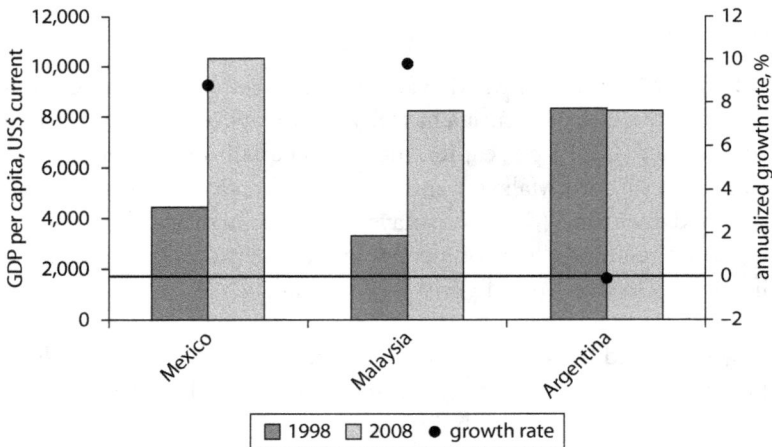

Source: World Bank, World Development Indicators.

Figure P4.2   Gini Index, Group D Countries, 1998–2008

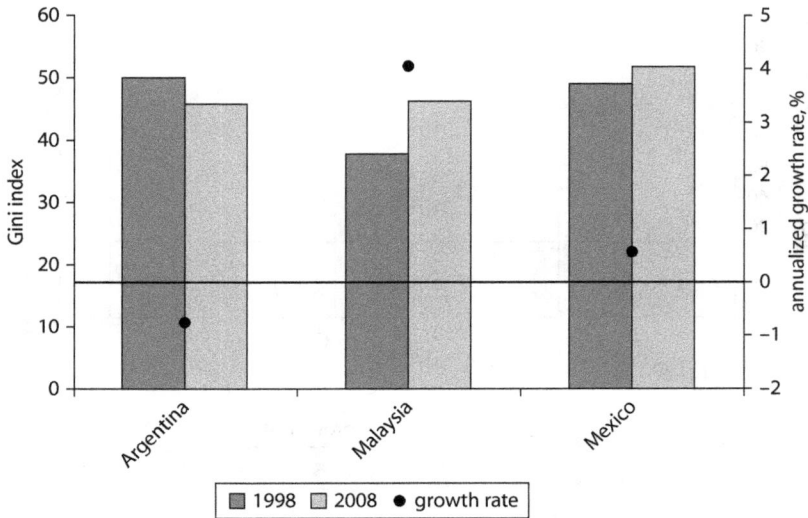

Source: World Bank, World Development Indicators.
Note: The Gini index measures the extent to which the distribution of income (or consumption expenditure) among individuals or households within an economy deviates from a perfectly equal distribution. A Gini index of 0 represents perfect equality, while an index of 100 implies perfect inequality (World Bank, *World Development Indicators*).

Fiscal Indicators for Group D

Figure P4.3   General Government Net Lending or Borrowing, Group D Countries, 1998–2008

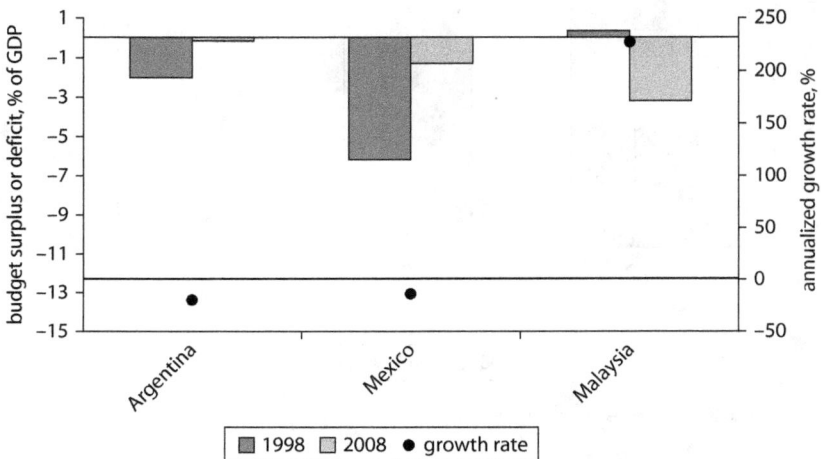

Source: IMF reports, various years.

Figure P4.4   General Government Gross Debt, Group D Countries, 1998–2008

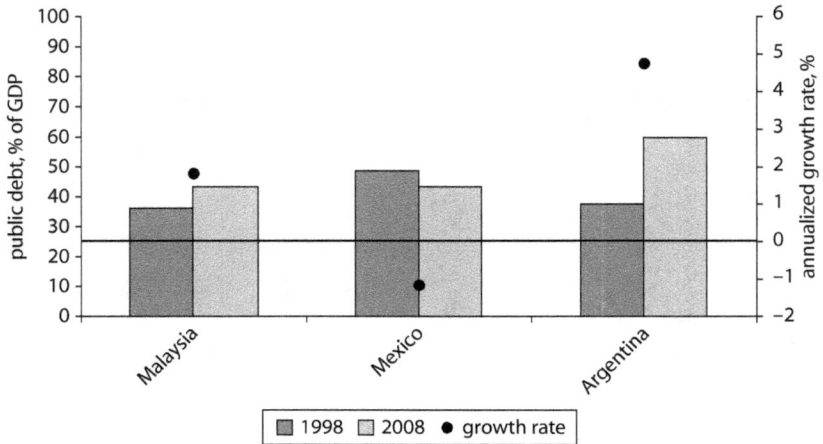

Source: IMF reports, various years.

Fossil Fuel Dependence for Group D

Figure P4.5   Electricity Production from Fossil Fuels, Group D Countries, 1998–2008

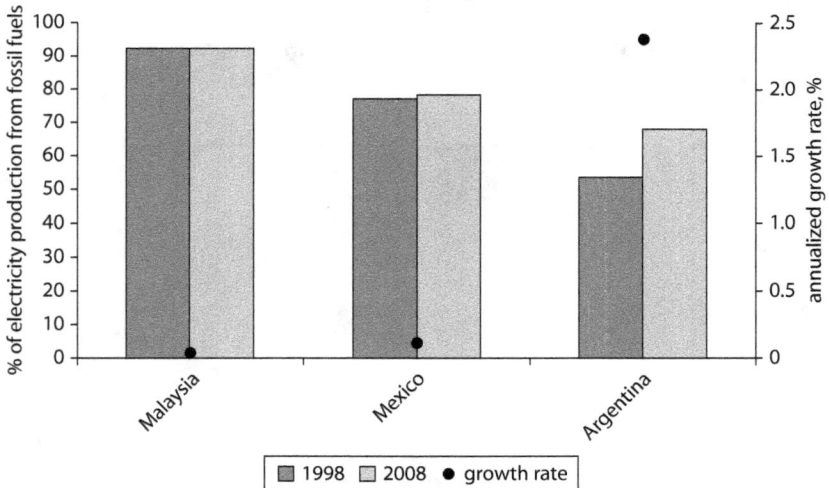

Source: World Bank, World Development Indicators.

Figure P4.6   Energy Net Imports, Group D Countries, 1998–2008

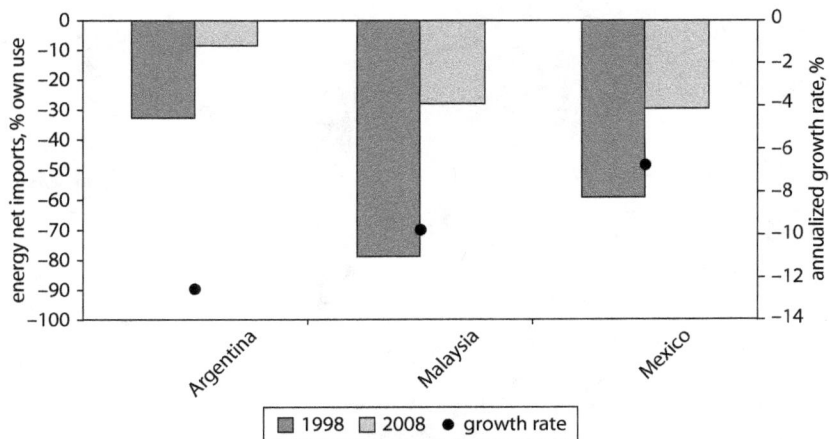

Source: World Bank, World Development Indicators.

CHAPTER 18

Argentina

Incentives to Energy Subsidy Reforms

Electricity and Natural Gas

As Latin America's largest natural gas producer, Argentina has a regulatory framework and tariff structure that has been designed to create incentives for oil producers to sell natural gas in the domestic market at a price below that of the world market. Below-market tariff freezes have constrained the private investments in generation and transmission capacity needed to keep pace with growing demand, forcing the government to take a leading role in driving forward new projects, including two new combined-cycle power plants, a new nuclear power plant, and an increase in hydroelectric output.

Energy subsidies increased from a low level of 0.1 percent in the early 2000s to 1.7 percent in 2010 (see figure 18A.1). The government has been increasing funding to the state energy company Enarsa and the wholesale power regulator, *Compañía Administradora del Mercado Mayorista Eléctrico* (CAMMESA). CAMMESA reportedly owes billions of dollars to energy companies, to whom it pays subsidies each year. Argentina froze most oil, gas, and electricity rates across the board following the 2001–02 economic crisis. To keep rates low, the government has been setting aside a substantial amount of subsidies every year to compensate the companies for lost income. But in recent years, operating costs have soared and demand for electricity has surged amid booming economic growth, making it more expensive for the government to keep prices frozen.

Cold winters led demand for natural gas and electricity to break records, pushing the government to ration gas to hundreds of industrial companies so it could guarantee the supply of gas to residential customers. Demand continues to be constrained by supply shortages. Underinvestment led to cuts in the supply of natural gas to industries and power plants in the recent past. There have also been shortages of diesel oil as a result of growing consumption, coupled with artificially low domestic prices. This has particularly affected the agricultural sector, for which diesel oil is a key input.

The slow postcrisis economic recovery can, to some extent, be understood and explained by the tariff freezing, which, as an effectively nondirect subsidy, has given industry and consumers an unfair advantage in the form of cheap economic "inputs." At the same time, market critics point out that the use of state subsidies to keep prices below market levels is tantamount to providing profit protection to the private sector.

The market for liquefied petroleum gas (LPG) is controlled by a distributing oligopoly that keeps prices high despite the fact that the government subsidizes this fuel. As a consequence of this situation, many low-income consumers who lack access to or cannot afford electricity use bottled gas designated for low-income consumers. The reason why the 10 kilogram (kg) LPG cylinders are not available for poor consumers at the subsidized price is that price ceilings were not set for retail sales. Since 2001, the price of a 10 kg refill of an LPG cylinder has increased fourfold, whereas the government has raised modestly the price of natural gas piped directly into the home, which is predominantly used in middle- and upper-income neighborhoods. In the four northeastern provinces, where no natural gas pipelines exist, consumers heavily rely on LPG for cooking and heating.

Reform Efforts

Argentina provides one of the best examples of full-scale power market reform in the world to date. Since the start of reforms in the late 1980s, Argentina's electricity industry went through comprehensive changes in the 1990s that involved the unbundling and privatization of the integrated state companies and the creation of a competitive wholesale electricity market.

The 1992 Electricity Regulation Act[1] was the keystone for the ambitious reform and privatization of the sector. It established the legal framework for further restructuring and privatization intended to stimulate competition and to benefit consumers in terms of both price and quality of service. It provided for the privatization of virtually all commercial activities that had been carried out by federally owned enterprises. It established the basis for the independent sector regulator (*Ente Nacional Regulador de la Electricidad*; ENRE) and other institutional authorities in the sector, and it created the Wholesale Electricity Market (*Mercado Eléctrico Mayorista*; MEM)[2] and its independent operator, CAMMESA. It also laid the basis for the administration of the wholesale power market, pricing in the spot market, tariff setting in regulated areas, and the valuation of the assets to be privatized.

Argentina's electricity market is now characterized by numerous producers in a highly competitive generation market. Many companies have more than 1,000 megawatts (MW) of installed capacity, and the largest one has 1,400 MW, which constitutes only 6 percent of total installed capacity. The initial generation market thus has had a low concentration ratio.

The three distribution companies divested from the former *Servicios Eléctricos del Gran Buenos Aires* (SEGBA)—Edenor, Edesur, and Edelap[3]— represent 44 percent of the electricity market in Argentina. Including the companies divested from some regional utilities, private participation in the distribution market has increased to 60 percent. The remaining distribution companies have remained in the hands of the provincial governments, but this ownership structure will change with the expansion of the new regulatory framework to the different regions of the country.

MEM wholesale prices fell significantly because of the installation of increasingly efficient capacity and the improving operating efficiencies of former state enterprises under competitive pressures. The price declined from US$45 per megawatt-hour (MWh) in 1992 to about US$25 per MWh in 1998; increased to US$27 in 2000; dropped to around US$10 per MWh in the depths of the economic crisis in 2002; and subsequently increased steadily to between US$35 and US$40 per MWh in 2007. When the power system had overcapacity, the resulting low prices in the short term caused most transactions to occur in the spot market. Competition was helped by the large margins between total installed generation capacity in the market and the demand for power. Part of this margin was needed, however, to cover the drop in hydropower output during dry seasons (Vagliasindi and Besant-Jones, forthcoming).

The reforms that had been implemented in Argentina were therefore deemed to have been a success. However, in the wake of the macroeconomic crisis and the devaluation of the peso, generators and gas suppliers found themselves in critical financial difficulties, which caused a halt to additional investment. The current challenge facing reforms in the Argentine power sector is to rebound from the financial ramifications of peso devaluation and to promote private investment.

In September 2008, the government announced the first rise in consumer tariffs since 2001. For residential consumers, the tariff increase ranged between 10 percent and 30 percent, according to the level of consumption (although low-consumption households were exempted from this rise). The rise reached 10 percent for commercial users and ranged between 10 percent and 15 percent for industrial users. In November 2008, the government implemented another tariff increase for residential and industrial consumers.

In August 2009, under pressure from unions and Congress, the government noted that the economic downturn had reduced households' purchasing power and that some consumers had been hit harder than expected by the higher tariffs. The government announced that tariff rises would be suspended for the peak southern hemisphere winter months of June and July, at a fiscal cost of close to US$130 million for the two months.

Poverty Alleviation Measures

Evidence from Household Surveys

Argentina has a long tradition of subsidizing energy tariffs for low-income pensioners, a practice that was preserved through the privatization process of the early 1990s. Eligibility for these subsidies was confined to low-income pensioners whose energy consumption did not exceed certain threshold levels. Until 1997, this subsidy took the form of a 50 percent discount on the fixed charge and volumetric charge for consumption of up to 210 kilowatt-hours (kWh) per month of electricity in the metropolitan area. In the case of natural gas, there was an 80 percent discount on the fixed charge and a tapered subsidy up to a consumption of 250 cubic meters per month—which was, overall, equivalent to a 50 percent subsidy on the consumption of a typical pensioner household in the metropolitan area. A direct government transfer to the respective utilities covered the cost of these discounts. In 1997, the discounts were phased out and replaced by a monthly transfer of US$13.50 to low-income pensioners or US$24 to pensioners using natural gas in the Patagonian region.[4]

In addition, a national Tariff Compensation Fund has been established for some years in the electricity sector. It is financed from 60 percent of the revenues generated by a surcharge of US$0.024 per kWh on all electricity traded through the national wholesale market, which amounted to US$98 million in 2002. The Federal Electricity Council distributes these resources to the provinces on the basis of a formula that seeks to compensate for differences in the cost of electricity production across jurisdictions. The underlying principle is one of horizontal equity, which seeks to equalize the electricity tariff across the country.

Argentina also uses province-level, means-tested subsidies for water and electricity. The electricity and water programs are on average progressive (see figure 18A.9), but the bulk of poor households are excluded from receiving subsidies in both cases.

Social Safety Nets

Argentina established the social protection program *Jefes y Jefas* in January 2002 in response to the economic and political crisis that hit Argentina at the end of 2001. *Jefes* was designed as an emergency program with a strategy of employment generation. The state participated as employer of last resort. The program has a decentralized administration and operations but is consolidated into a centralized national database.

In 2005, the government started implementing a transition strategy to phase out *Jefes* and move to two new programs that are part of a long-term social protection strategy: (a) an employment benefit and training program, *Seguro*, and (b) a conditional cash transfer (CCT) program, *Familias*. The transition was expected to be completed by 2011 with the goal of transferring all *Jefes* beneficiaries to one of the two programs. By the end of 2008, *Familias* was approaching 600,000 beneficiaries (World Bank 2011a). Payments are generally transferred through debit cards with the *Banco de la Nación Argentina*. Recent studies show that the social safety program seems to be relatively well targeted, with more than 70 percent of beneficiaries belonging to the poorest 25 percent of the population (Alperin 2009). Further developing the targeting and effectiveness of the CCT program will be essential for compensating the most affected households from tariff increases as Argentina phases out subsidies.

Using a dynamic computable general equilibrium model for Argentina, Benitez and Chisari (2010) simulate the economic and social impact of removing residential consumer electricity and natural gas subsidies. The model uses a social accounting matrix for the Argentinean economy in 2006, when household subsidies accounted for around 0.4 percent of gross domestic product. Figure 18A.10 reports the results of the welfare impact of subsidy removal as a loss in consumer real income under two scenarios: (a) without any mitigating policy to protect the poor, and (b) with cash transfers targeted to the poorest decile of the population. All incremental changes are shown by percentage for the 2007–11 period with respect to the benchmark. The results show how cash transfers mitigate the social impact of subsidy removal on the poor, with substantial incremental increases in real income for all periods. It is also worth noting that in the absence of mitigating policies, welfare losses are experienced by the poorest decile only during the first year following the reforms.

Key Lessons Learned

The reform in Argentina can be seen as moderately successful prior to the collapse of the Argentine peso in early 2002. Tariff reforms in 2008 did not achieve their full effect because of the impacts of the global financial crisis and political considerations that have forced the government to backtrack, at least temporarily, on some of the planned price increases.

The existence of well-targeted social safety programs will provide Argentina with strong foundations for implementation of subsidy reforms while protecting the poor.

Annex 18.1 Argentina Case Study Figures

INCOME LEVEL: Upper-middle income
REGION: Latin America and the Caribbean
ENERGY NET IMPORTER/EXPORTER: Net exporter
SUBSIDIES: Electricity, LPG, natural gas
PHASING OUT SUBSIDIES: Ongoing

Fiscal Burden of Energy Subsidy in Argentina

Figure 18A.1   Explicit Budgetary Energy Subsidies in Argentina, 2000–10

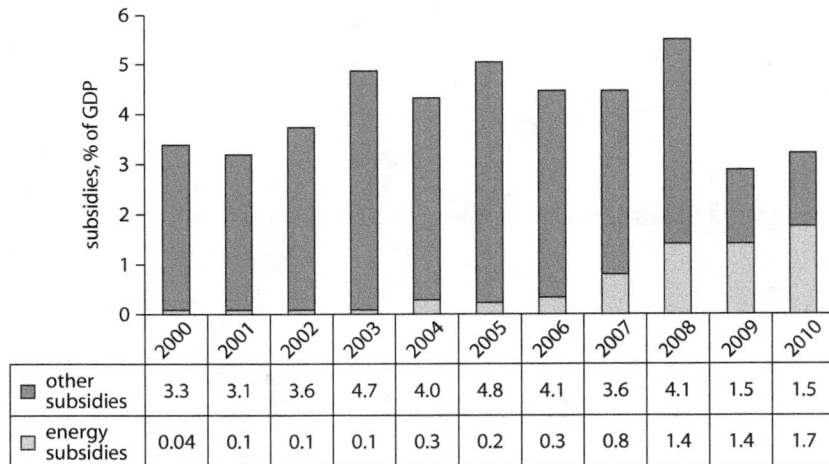

	2000	2001	2002	2003	2004	2005	2006	2007	2008	2009	2010
▣ other subsidies	3.3	3.1	3.6	4.7	4.0	4.8	4.1	3.6	4.1	1.5	1.5
▢ energy subsidies	0.04	0.1	0.1	0.1	0.3	0.2	0.3	0.8	1.4	1.4	1.7

Source: Asociación Argentina de Presupuesto y Administración Financiera Pública, various years.

Fuel Prices and Road Sector Consumption in Argentina

Figure 18A.2   Domestic Retail Fuel Prices in Argentina, 2002–10

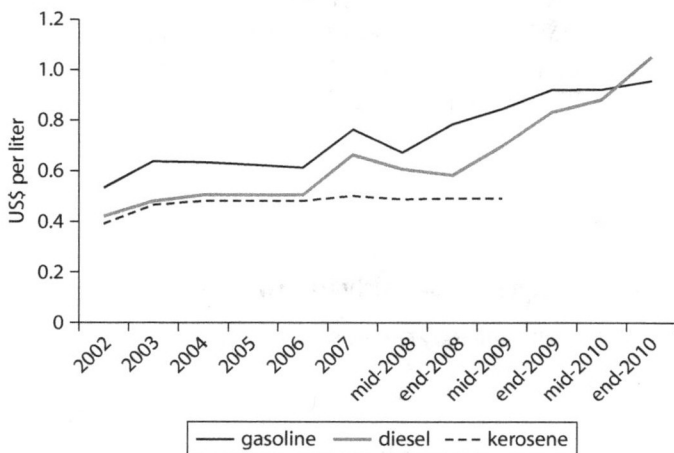

Source: IMF reports and GIZ n.d., various years.

Figure 18A.3   Road Sector Diesel Consumption in Argentina, 1998–2008

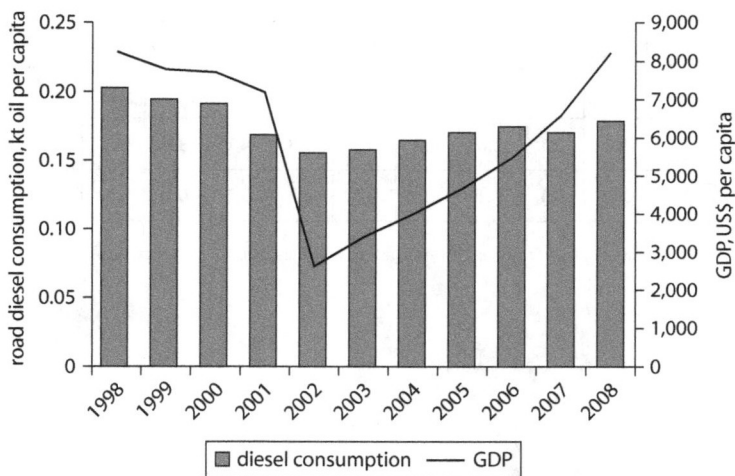

Source: World Bank, World Development Indicators.

Figure 18A.4   Road Sector Gasoline Consumption in Argentina, 1998–2008

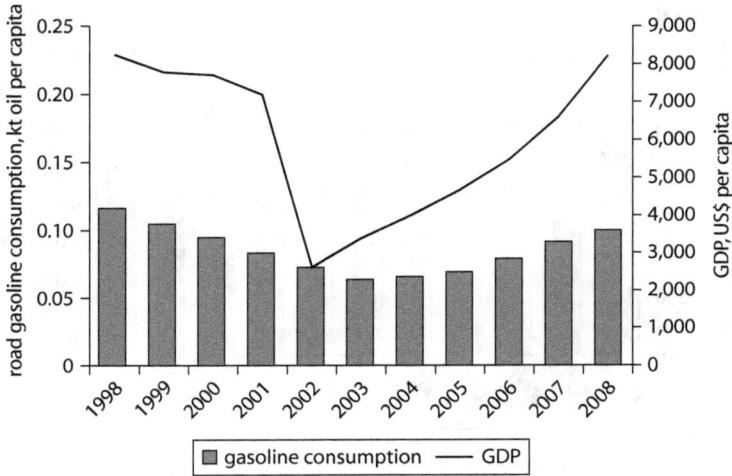

Source: World Bank, World Development Indicators.

Electricity Price and Power Consumption in Argentina

Figure 18A.5   Average Electricity Price in Argentina, 1998–2009

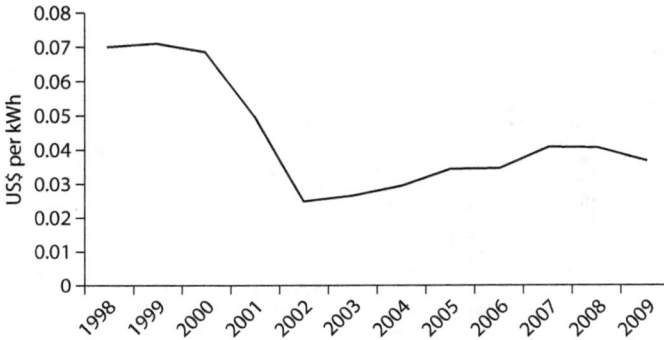

Sources: Vagliasindi and Besant-Jones, forthcoming; Power Market Structure Database.
Note: kWh = kilowatt-hour.

Figure 18A.6   Power Consumption Per Capita in Argentina, 1998–2008

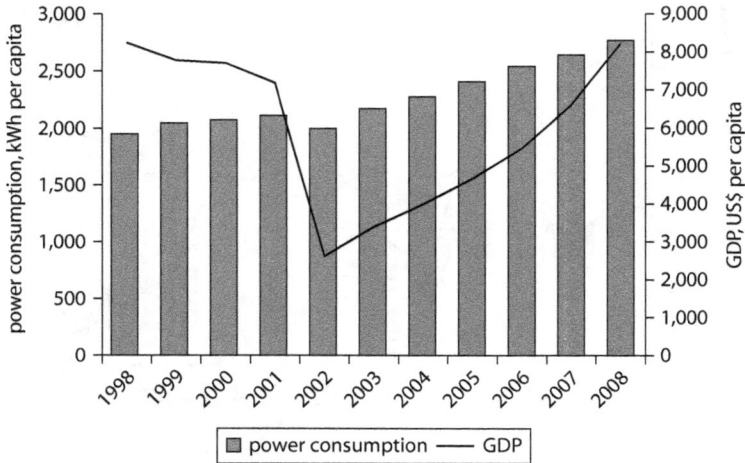

Source: World Bank, World Development Indicators.
Note: kWh = kilowatt-hour.

Poverty Impact Evidence from Household Surveys in Argentina

Figure 18A.7   Electricity Block Tariffs in Argentina, 2010

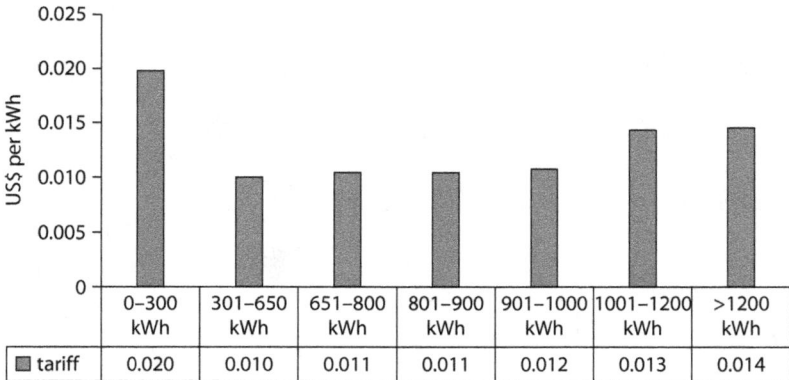

	0–300 kWh	301–650 kWh	651–800 kWh	801–900 kWh	901–1000 kWh	1001–1200 kWh	>1200 kWh
▣ tariff	0.020	0.010	0.011	0.011	0.012	0.013	0.014

Sources: Vagliasindi and Besant-Jones, forthcoming; Power Market Structure Database.
Note: kWh = kilowatt-hour.

Figure 18A.8   Household Electricity Expenditure in Argentina, by Income Quintile, 2002

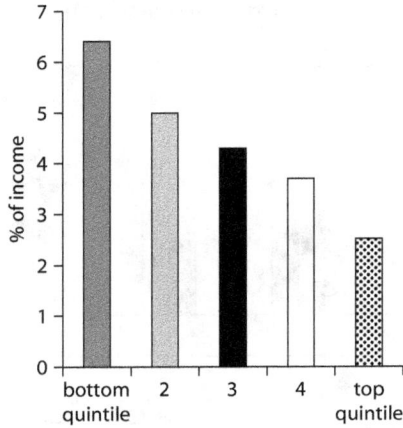

Source: Foster 2004.

Figure 18A.9   Benefit Incidence of Means-Tested Subsidies in Argentina, by Income Quintile, 2002

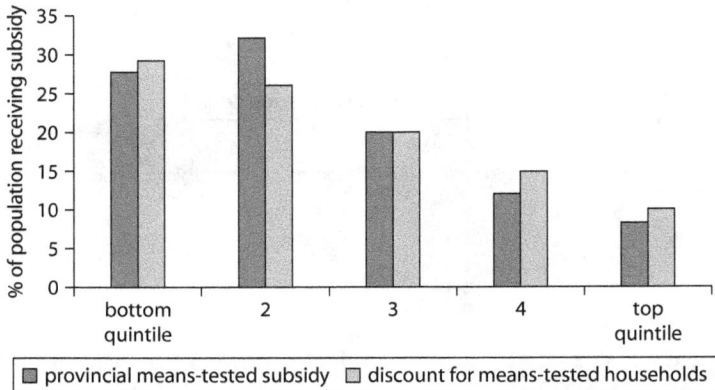

Source: Foster 2004.

Figure 18A.10   Welfare Impact of Removal of Electricity and Natural Gas Subsidies in Argentina, by Income Decile, 2007–2011

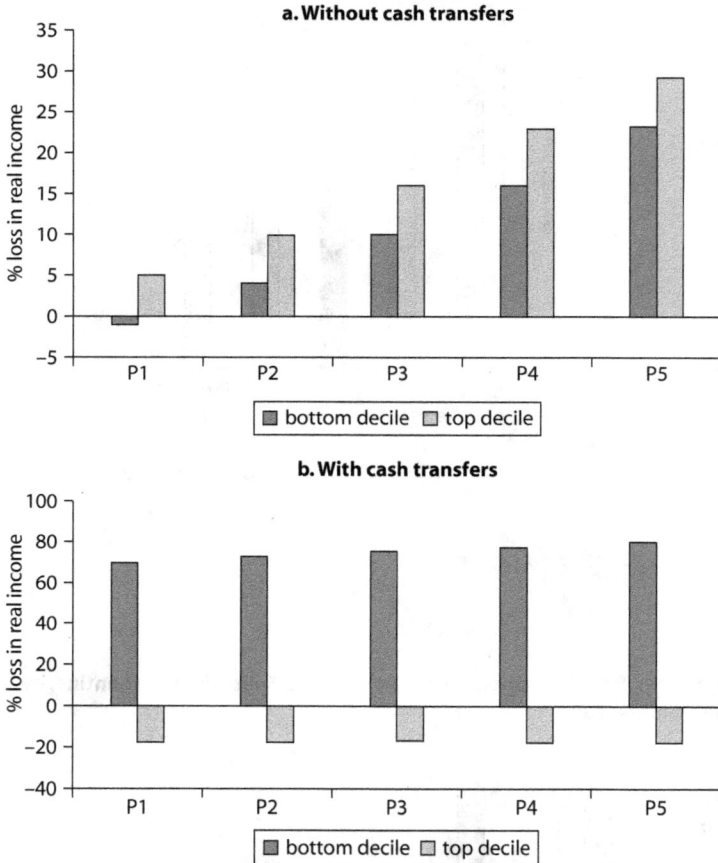

a. Without cash transfers

bottom decile   top decile

b. With cash transfers

bottom decile   top decile

Source: Benitez and Chisari 2010.
Note: P1–P5 refers to the five-year period 2007–11 after the simulation of the tariff increase in 2006.

Notes

1. Law No. 24065 (1992).

2. Resolution 38/91, SEE, established rules for MEM (World Bank 2011b).

3. Edenor was incorporated under the name *Empresa Distribuidora Norte Sociedad Anónima;* Edesur under the name *Electricidad Distribuidora Sur;* and Edelap under the name *Empresa de Electricidad de la Plata.*

4. As a result of Decree 319/97.

References

Alperin, María Noel Pi. 2009. "The Impact of Argentina's Social Assistance Program Plan *Jefes y Jefas de Hogar* on Structural Poverty." *Estudios Económicos* special issue (2009): 49-81.

Benitez, Daniel, and Omar Chisari. 2010. "Recycling Universal Energy Subsidies on Residential Consumers: Alternative Fiscal Policies to Help the Poorest." Unpublished manuscript, World Bank, Washington, DC.

Foster, Vivien. 2004. "Toward a Social Policy for Argentina's Infrastructure Sectors: Evaluating the Past, Exploring the Future." Policy Research Working Paper 3422, World Bank, Washington, DC.

GIZ (German Agency for International Cooperation). n.d. International Fuel Prices database. GIZ (formerly GTZ), Bonn. http://www.gtz.de/en/themen/29957.htm.

Vagliasindi, Maria, and John Besant-Jones. Forthcoming. *Power Market Structure: Revisiting Policy Options.* Directions in Development Series. Washington, DC: World Bank.

World Bank. 2011a. "Argentina Basic Protection Project." Human Development Sector; Argentina, Chile, Paraguay, and Uruguay Country Management Unit; Latin America and the Caribbean Region, World Bank, Washington, DC. http://web.worldbank.org/external/projects/main?Projectid=P115183&theSitePK=40941&pagePK=64283627&menuPK=228424&piPK=73230.

———. 2011b. "Revisiting Policy Options on the Market Structure in the Power Sector." Report of the Energy Sector Management Assistance Program (ESMAP) with the Public-Private Infrastructure Advisory Facility (PPIAF), World Bank, Washington, DC.

CHAPTER 19

Malaysia

Incentives to Energy Subsidy Reforms

The energy sector is a major contributor to the Malaysian economy, with liquefied natural gas (LNG) and crude oil together accounting for 10.2 percent of total export revenues in 2009. That same year, Malaysia was the second-largest exporter of LNG in the world, after Qatar. Malaysia LNG, a subsidiary of the state-owned oil company Petroliam Nasional (Petronas), has increased its customer base of Japanese power and gas companies to 13 firms in recent years, and in several cases is the sole supplier to these companies. Malaysia also supplies LNG to China under a 25-year contract that commenced in 2009. Owing to distortions in the domestic energy market, including pricing, demand for petroleum products has been growing faster than production. There is a possibility that Malaysia could become a net importer of petroleum products toward 2020.

The main pressures to reduce subsidies were (a) the growing subsidy bill during the oil price boom, which reached a peak in early July 2008 and amounted to 2.4 percent of the Malaysian gross domestic product (GDP) (see figure 19A.1); and (b) the high fiscal deficit, which in 2009 reached 7 percent of GDP. The budget deficit also grew as a result of economic stimulus measures, which—when combined with subsidies—drove government expenditures to new heights. Excluding those for education and health, subsidies account for some 20 percent of government spending, or about 5 percent of GDP, with the bulk going to reducing the cost of fuel to consumers and industry (IMF 2010). On the expenditure side, the authorities are considering overhauling Malaysia's fragmented and badly targeted social safety nets.

The subsidy rationalization program is part of the broader reform agenda to remove distortions in the economy, improve competitiveness and market efficiency, and ensure a more optimal use of scarce resources. The existing social safety net is also being reviewed to achieve better targeting and effectiveness. The rollout of the long-delayed goods and services tax (GST) has been announced, but legislative details are still in the works. The GST is intended to broaden the tax base by replacing two existing taxes in a revenue-neutral fashion (IMF 2010).

Reform Efforts

Fossil Fuels

The most significant subsidy reform Malaysia undertook was in early July 2008, at the peak of the high international oil prices, when, in an effort to cut the subsidy bill, gasoline prices increased by 40 percent and diesel by 63 percent (IEA 2009). To offset the increased prices, the Malaysian government offered cash rebates in the form of lower annual road taxes. Other than subsidy reductions and cash rebates, the package included windfall taxation on certain sectors and an expansion of the social safety net (IEA 2009). With the dramatic drop in oil prices in the second half of 2008, it became easier for Malaysia to further reduce its gasoline subsidies because prices were declining. From August to November 2008, fuel prices were reduced five times (see figure 19A.2).

In July 2010, the government reduced subsidies for some key commodities. Low-octane gasoline and diesel prices have been increased by some 3 percent, and the price of liquefied petroleum gas has gone up by about 6 percent. High octane gasoline is no longer subsidized. Although these measures fall short of earlier official proposals, they mark the beginning of the subsidy reform program highlighted in Malaysian Prime Minister Najib's New Economic Model (IMF 2010).

Electricity

Tenaga Nasional (TNB) is the largest state-owned power company, accounting for more than 60 percent of generating capacity in peninsular

Malaysia. TNB is involved in the generation, transmission, and distribution of electricity. TNB was constrained by the obligation to pay Malaysia's nine independent power producers (IPPs) irrespective of the level of power demand under rigid power purchase agreements. The IPPs hold 21-year concessions that will end in 2015 and that contribute around 40 percent of national electricity supply.

In August 2008, the price of gas for power generation was raised by 124 percent in peninsular Malaysia, and the average electricity tariff for all sectors of the economy was increased by 24 percent (from US$0.075 per kilowatt-hour [kWh] to US$0.093 per kWh), in line with the increase in the gas price (IEA 2009). The prices of fuel were revised upward in December 2010.

Power consumption has increased steadily as a result of healthy economic growth. Malaysia has one of the highest levels of energy consumption per head among the countries that make up the Association of Southeast Asian Nations, at an estimated 2,710 kilogram oil equivalent in 2010. This is also reflected in the high level of power consumption per capita (see figure 19A.6).

Poverty Alleviation Measures

Evidence from Household Surveys

The pattern of household expenditure over time did not significantly change (see figure 19A.8). Between the years 1999 and 2005, the composition of fuel consumption of Malaysian households shows that petroleum fuels accounted for more than 60 percent of per capita household expenditure on energy sources. Thirty percent of that expenditure went toward electricity, with only around 10 percent spent on natural gas.

The trend over time shows that the petroleum fuels' share in household expenditure (as a percentage of income) increased sharply, from an average 5.2 percent in 1999 to 8.8 percent in 2005 (see figure 19A.9). The burden of household expenditure for electricity increased from 2.3 percent of income to about 3.6 percent over the same period. Inequality in fuel expenditure for Malaysian households also increased over time, with the energy Gini coefficient increasing from 0.41 to 0.49 from 1999 to 2005 (Moradkhani et al. 2010).

Moradkhani et al. (2010) model the direct welfare effects of doubling the prices of all sources of energy. For petroleum fuels, the most adversely affected groups are the top quintiles, whose loss of real income is above 9 percent. This is more than twice the loss of real income of the lowest quintile, which was estimated at around 4 percent. No significant disparity of real income loss was found for electricity among the quintiles: all lost about 3 percent. In the case of natural gas, the loss of real income for the bottom quintile amounts to 1.1 percent, more than 5 times the real loss of income for the top quintile (see figure 19A.11).

Social Safety Nets

Malaysia lacks comprehensive social safety nets and targeted measures for the purpose of mitigating the impact of higher prices on the poor (IMF 2009). The government has been making efforts to enhance its social targeting programs by recently setting up a centralized system, e-Kasih, for identification of poor households and for managing a list of potential beneficiaries. The system suffers from a number of problems such as coverage and accuracy issues related to the existing databases and new household visits. It is also unclear whether and how the system should capture beneficiaries of safety net programs that are based on criteria other than income. More important, perhaps, the measures taken to date fail to address some of the more fundamental challenges associated with the current targeting system, in particular the focus on cash income and the inability to verify income for many households.

Overall, the social safety net system in Malaysia is poorly targeted and fragmented across several government agencies (and levels of government). There will be a need to assess the extent to which safety net programs are achieving their intended objectives as well as broader aspects of program performance (for example, incentive effects and administrative costs). There is also a need to assess targeting performance. This would include looking at errors of inclusion and exclusion, the direct and indirect costs of targeting, and the potential merits of alternative targeting approaches, such as proxy means testing or community-based targeting (World Bank 2009). The establishment of the e-Kasih mechanism is a step in the right direction, nevertheless.

To mitigate the impact on the population and prevent social unrest when prices increased in 2008, the government started to issue smart cards to owners of motor vehicles below a certain engine size. The policy was adopted because most Malaysians own cars with relatively small engines from local car manufacturers. Smart cards have proven to be an effective way of transferring targeted subsidies.

Malaysia's experience is quite instructive. To contain the subsidy bill, Malaysia launched a smart card scheme in 2006 for two consumer categories: public transport operators and fishermen. This followed the failure of a previous attempt (January 2005) to contain the subsidy by limiting diesel fuel supplies to filling stations based on 2003 sales. That quota system led to diesel fuel shortages, and a quarter of the filling stations reportedly ran out of diesel fuel. Under the smart card system, subsidized fuels are rationed, with the monthly quota based on the vehicle category or boat size (World Bank 2009).

The card is called *MyKad* in Malaysia. *MyKad* consolidates drivers' licenses and identification cards for bill payment (ePurse), tolls, parking for public transport, automated teller machine banking, and health services. Citizen acceptance was de facto as the government began to automatically issue *MyKad* to all citizens who reach the age of 12 and to any who have lost their old national identification cards. The uptake has been relatively slow. In May 2006, the government of Malaysia awarded a US$5 million contract to a consortium of companies to roll out a new stage of the project and enable all of Malaysia's 25 million inhabitants to benefit from the smart cards.

Key Lessons Learned

Malaysia has yet to completely phase out fuel subsidies, but it has taken important steps to reduce them and is further developing its social safety nets to reduce the adverse impact on consumers.

Malaysia is using smart card technology to facilitate fuel rationing. The cards identify beneficiaries according to different categories of motorized vehicles. To achieve more productive uses of its remaining energy potential, Malaysia is prioritizing energy efficiency initiatives in its comprehensive National Energy Plan. The impact of the reforms has also encouraged businesses with large utility bills to seek more energy-efficient solutions.

Annex 19.1 Malaysia Case Study Figures

INCOME LEVEL: Upper-middle income
REGION: East Asia and Pacific
ENERGY NET IMPORTER/EXPORTER: Net exporter
SUBSIDIES: Gasoline, diesel
PHASING OUT SUBSIDIES: Ongoing

Fiscal Burden of Energy Subsidy in Malaysia

Figure 19A.1   Explicit Budgetary Energy Subsidies in Malaysia, 2004–10

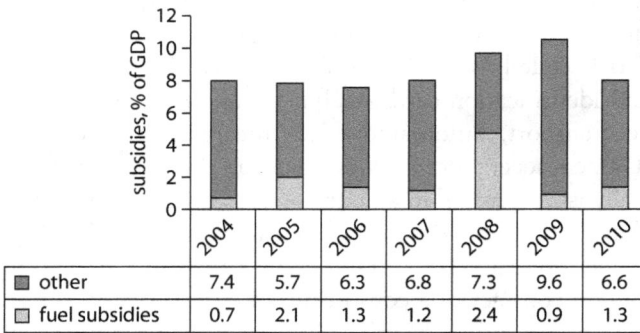

	2004	2005	2006	2007	2008	2009	2010
▨ other	7.4	5.7	6.3	6.8	7.3	9.6	6.6
☐ fuel subsidies	0.7	2.1	1.3	1.2	2.4	0.9	1.3

Sources: IMF 2009 and other Article IV staff consultations, various years.

Fuel Prices and Road Sector Consumption in Malaysia

Figure 19A.2   Domestic Retail Fuel Prices in Malaysia, 2002–10

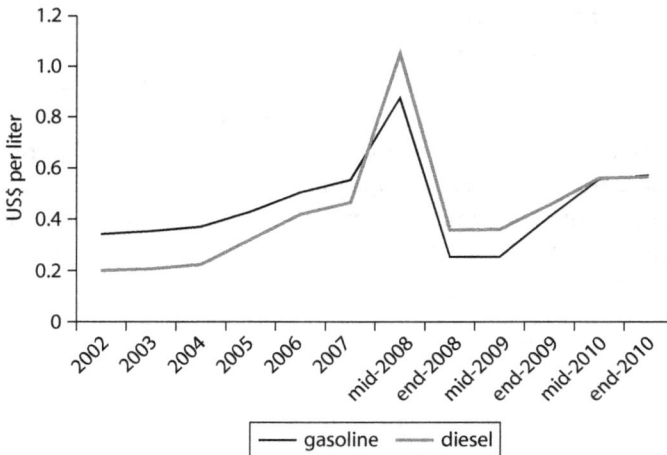

Sources: GIZ n.d.; Malaysia Department of Statistics, various years.

Figure 19A.3   Road Sector Diesel Consumption in Malaysia, 1998–2008

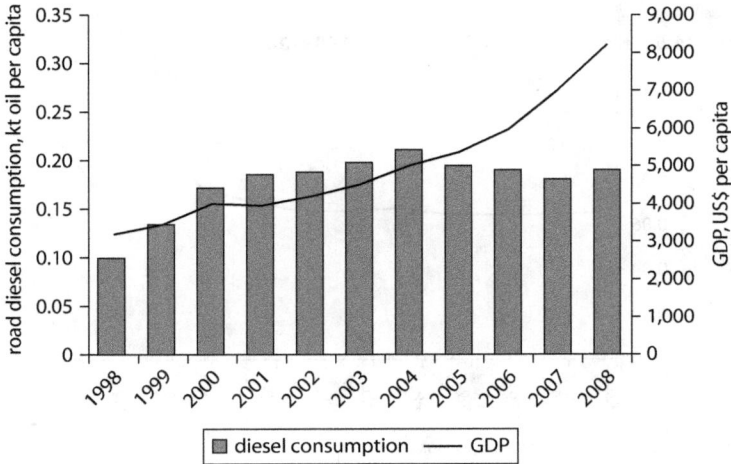

Source: World Bank, World Development Indicators.

Figure 19A.4   Road Sector Gasoline Consumption in Malaysia, 1998–2008

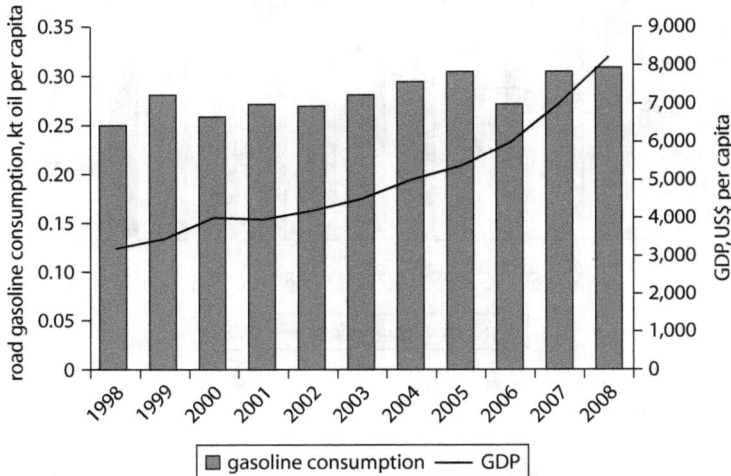

Source: World Bank, World Development Indicators.

Electricity Price and Power Consumption in Malaysia

Figure 19A.5   Electricity Price in Malaysia, 1999–2010

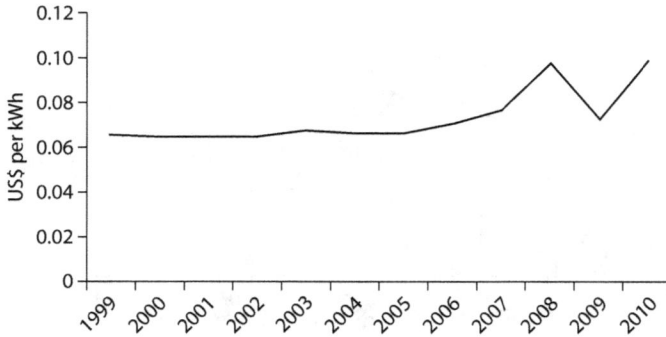

Source: Malaysia Department of Statistics, various years.
Note: kWh = kilowatt-hour.

Figure 19A.6   Power Consumption Per Capita in Malaysia, 1998–2008

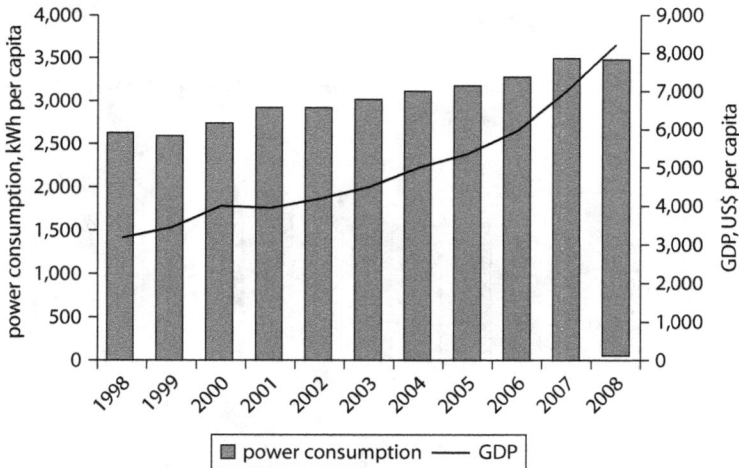

Source: World Bank, World Development Indicators.
Note: kWh = kilowatt-hour.

Poverty Impact Evidence from Household Surveys in Malaysia

Figure 19A.7   Electricity Block Tariffs in Malaysia, 2010

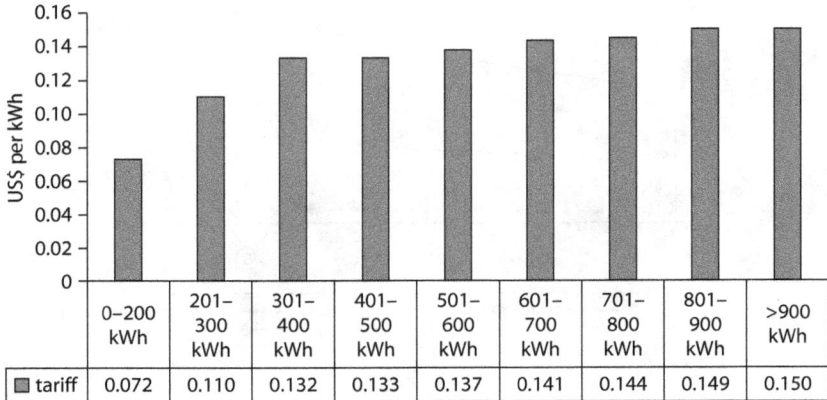

	0–200 kWh	201–300 kWh	301–400 kWh	401–500 kWh	501–600 kWh	601–700 kWh	701–800 kWh	801–900 kWh	>900 kWh
tariff	0.072	0.110	0.132	0.133	0.137	0.141	0.144	0.149	0.150

Source: TNB annual reports.
Note: kWh = kilowatt-hour.

Figure 19A.8   Per Capita Household Energy Expenditure in Malaysia, by Source, 1999 and 2005

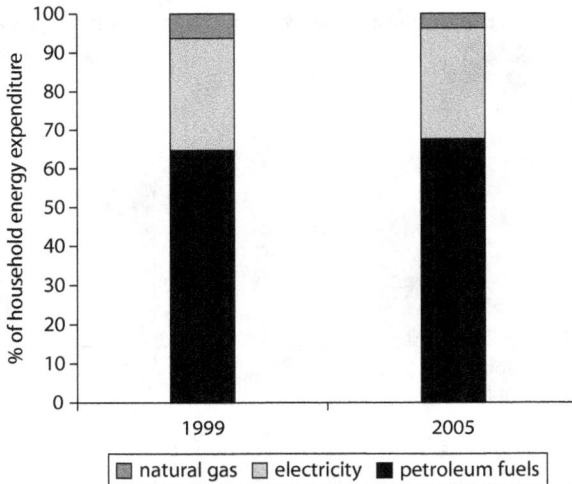

Source: Household expenditure surveys, 1999 and 2005.

Figure 19A.9   Average Household Energy Expenditure in Malaysia, 1999–2005

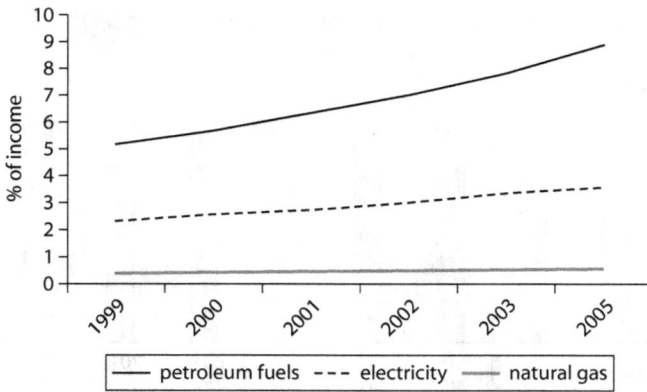

Source: Household expenditure surveys, 1999 and 2005.

Figure 19A.10   Fuel Expenditure in Malaysia, by Income Quintile, 1999 and 2005

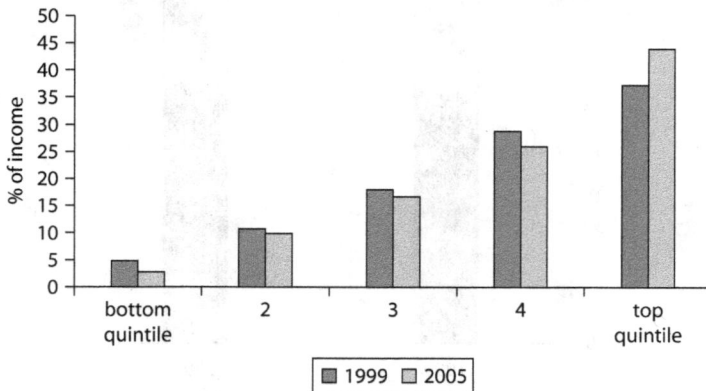

Source: Household expenditure surveys, 1999 and 2005.

Figure 19A.11   Welfare Impact of Phasing Out Subsidies in Malaysia, by Subsidy Source and Income Quintile

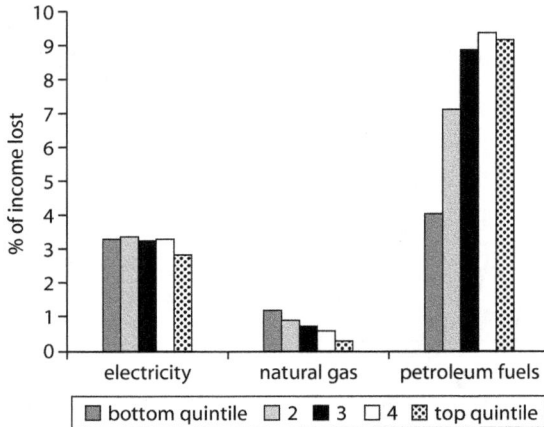

Source: Moradkhani et al. 2010, based on household expenditure survey, 2005.

References

GIZ (German Agency for International Cooperation). n.d. International Fuel Prices database. GIZ (formerly GTZ), Bonn. http://www.gtz.de/en/themen/29957.htm.

IEA (International Energy Agency). 2009. *World Energy Outlook 2009.* Paris: Organisation for Economic Co-operation and Development/IEA.

IMF (International Monetary Fund). 2009. "Malaysia: 2009 Article IV Consultation." Country Report 09/265, IMF, Washington, DC.

———. 2010. "Malaysia: 2010 Article IV Consultation." Country Report 10/265, IMF, Washington, DC.

Moradkhani, Narges, Zakariah Abd Rashid, Taufiq Hassan, and Anuuar Md Nassir. 2010. "The Impact of Increasing Energy Prices on the Prices of Other Goods and Household Expenditure: Evidence from Malaysia." Paper presented at the 18th International Conference of the International Input-Output Association (IIOA), Sydney, June 20–25.

World Bank. 2009. *Malaysia Economic Monitor,* November 2009. World Bank, Washington, DC. http://www.worldbank.org/en/country/Malaysia.

CHAPTER 20

Mexico

Incentives to Energy Subsidy Reforms

Mexico is the one of world's top producers of oil, though the share of oil production has been decreasing over time. Natural gas production has been increasing sharply but not enough to satisfy demand, so the country is a net importer of natural gas, mainly from the United States.[1]

The energy sector in Mexico is dominated by state-owned enterprises. The national state-owned company, Petróleos Mexicanos (PEMEX), enjoys a monopoly on production, refining, and distribution of oil products in the country. Although the Mexican government opened the downstream gas sector to private operators in 1995, PEMEX is currently the dominant operator in natural gas transmission. However, the private sector has acquired a significant role in natural gas distribution.

In the case of electricity—with the exception of generation, where significant entry occurred—electricity is provided by the state-owned Federal Electricity Commission (*Comisión Federal de Electricidad*; CFE), which is also responsible for planning, construction, and operation of the national electric system.

Fuel Subsidies

PEMEX is the single most important source of revenue for the Mexican government. In 2010, it accounted for 35 percent of total revenue. Oil products have a significant role in the energy matrix, accounting for 66 percent of total energy consumption in 2010. Gasoline and diesel are important oil products in the energy matrix. These two fuels are used heavily in Mexico for the transport and industry sectors (SIE n.d.).

Because Mexico lacks the capacity to produce all of the gasoline that the country demands, significant quantities (and higher-quality fuels) are imported from the United States.[2] Due to the recent gasoline price fluctuations, the government's subsidy burden increased by an amount that would have been enough to build several refineries to enhance the country's own petroleum fuels production capacity. In the past, Mexican gasoline prices were as much as 25 percent below U.S. prices, but they are slowly catching up since the government has started to scale back gasoline subsidies.

Electricity

Electricity subsidies in Mexico are reported by the utility companies as accounting costs resulting from below-cost pricing. However, the Mexican government essentially reimburses the CFE for providing subsidies to its customers by exempting it from the payment of taxes and dividends (*aprovechamiento*). Since 2002, the volume of subsidies has exceeded the notional amount of *aprovechamiento* and has therefore begun to erode the CFE's capital base. The federal government also used to provide the state-owned Luz y Fuerza del Centro a direct cash subsidy to cover its operating deficits and customer subsidies until the enterprise was closed in 2009, and its customers as well as its equipment and facilities have been taken over by the CFE.

Reform Efforts

Fossil Fuels

The prices of petroleum products (including diesel, gasoline, and liquefied petroleum gas) are determined on a monthly basis by the Mexican Ministry of Finance and the Energy Regulatory Commission. For gasoline and diesel, price-setting criteria include an international reference price in Houston, Texas; logistic adjustments; and taxes (the Impuesto Especial de Productos y Servicios [IEPS]—a special tax on services on gasoline and diesel—and the value added tax). The IEPS is a tax that buffers the difference between the retail price (set by the Ministry of Finance) and the producer price (determined by the international reference price plus logistic adjustments). For example, as long as the final price is higher than the producer price, an increase in the international reference price of gasoline increases the price to the producer and reduces the amount of the tax (OECD 2011).

The evolution of the international prices of petroleum and its derivatives in 2007 and 2008 caused a significant increase in the international prices of gasoline and diesel. Because of this, the IEPS was zero and the price charged to the public was, in some periods, below the international reference price, which translated into a subsidy on the sale of these products (Yepez-García, Luna-Tovar, and Portes 2012). Figure 20.1 shows that the largest revenue gap coming from the IEPS was recorded in 2008. As previously mentioned, the main factor behind the reduction of revenues is the significant price difference between the international reference price in Houston and the retail price in Mexico, which reduced the revenues that the government could have collected if the price of fuels had reflected international benchmarks. As a result of the price controls of fuels, during the 2007–08 period there was an implicit subsidy to consumers of around US$21.7 billion.

Figure 20.1   Expected vs. Actual Revenues from the IEPS, Mexico, 2007–11

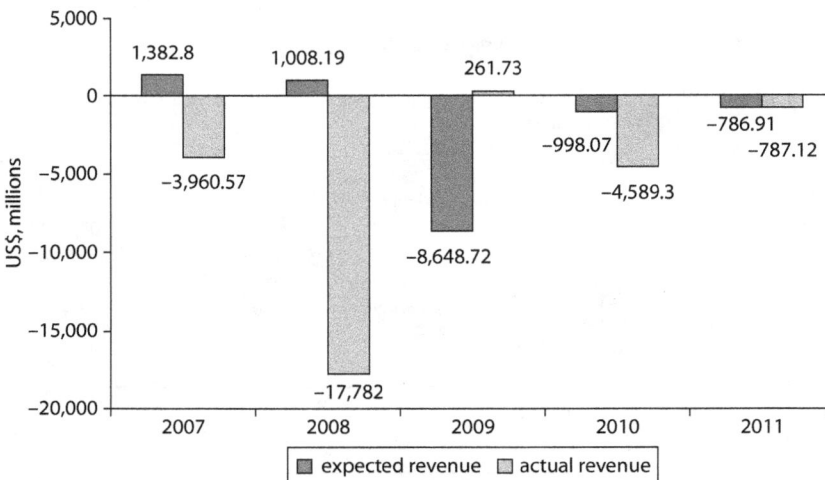

Due to the differential between imported gasoline prices and those in Mexico, PEMEX has forgone substantial revenues. The largest loss was during February–October 2008, when the price differential was the highest. During this period, the government implemented gradual adjustments of the prices of gasoline and diesel, taking into consideration the cost of inflation and the trend of the international reference price of gasoline in Houston. In 2009, the government froze the price of gasoline and adjusted the price of diesel, as figure 20.2 shows. From December 2009 on, the government adjusted the price of gasoline and allowed increases in the price of diesel.

A petroleum fund (*Fondo de Estabilización de los Ingresos Petroleros*; FEIP) was created in 2000 to smooth the impact of fluctuations in the price of oil on government revenues. As in Chile and Peru, it works in a countercyclical way, with the domestic sales price of petroleum products being either taxed or subsidized depending on variations in a benchmark import price. In contrast to Chile and Peru, where the funds use a predetermined formula, the Mexican fund relies on internal price forecasts that are largely influenced by the level of inflation.

The formula was not designed to absorb the impact of substantial increases in petroleum prices since 2007. Starting in 2007, the FEIP resulted in subsidies (reported as negative taxes in Mexico). About a quarter of oil revenues accruing in FEIP comes from net exports and vary directly with world prices. To protect these revenues from price

Figure 20.2   Gasoline and Diesel Prices in Mexico, 2007–11

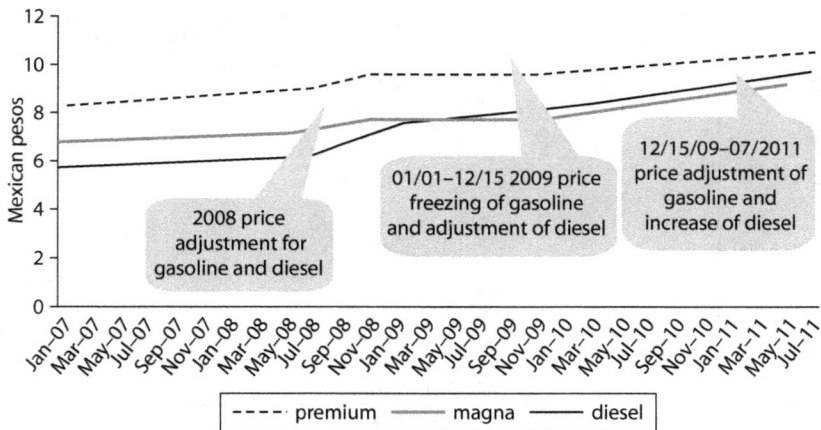

Source: Reyes Tepach, based on data from PEMEX 2011.

volatility, authorities use resources from the FEIP to purchase one-year put options with a strike price that is equivalent to the projected oil price for the fiscal budget. This "hedging" strategy proved to be particularly successful in 2009, when it protected the budget from the drop in oil prices.

Electricity

In Mexico, the electricity sector constitutes a legal monopoly. Generation, transmission, and distribution of electricity in Mexico are delegated to the CFE. In addition, Mexico has an array of private producers that generate electricity for their own use and independent power producers that generate electricity for the CFE under long-term agreements.

There are three types of residential rates:

- *Rate 1*: applicable year-round in areas with temperate climate
- *Rates 1A, 1B, 1C, 1D, 1E, and 1F*: applicable in areas with hot climates, depending on the average temperature over the past three years in the corresponding areas
- *Rate DAC*: applicable to high-consumption users who exceed a given average demand.

Electricity subsidies were first introduced in 1973 in the form of increasing block tariffs, with tariffs well below average costs for low consumption values (namely the first two blocks). Shortly after, Mexico started offering "summer subsidies" (that is, subsidized rates for customers living in areas where temperatures exceed 25 degrees Celsius for four months of the year). A tariff review in 2002 tried to reduce the amount of subsidies by charging higher tariffs to the consumers characterized by higher consumption, but 75 percent of total consumption still remained subsidized.

From 1980 to 1990, real average residential electricity prices decreased at an annual rate of 3.5 percent. They increased at a rate of 6.5 percent per year between 1990 and 2000 and then remained constant until 2006 (Rosas-Flores and Galvez 2010).

Poverty Alleviation Measures

Evidence from Household Surveys

Mexico's electricity tariff structure is quite complex because of the existence of several tariff categories, which are also differentiated by region and average temperature. In line with the findings of Komives et al. (2005), which reviewed the experience of increasing block tariffs, Komives et al. (2009) confirmed the highly regressive nature of residential electricity subsidies in Mexico, with residential electricity subsidies disproportionately benefiting the richest income deciles of customers, especially in the case of "summer subsidies." Among the customers belonging to Tariff 1F (the warmest of the climate-based tariffs from 1A to 1F), the top income decile received 20 percent of the total subsidies in 2005.

Analysis of data from the Mexican National Household Income and Expenditure Survey (ENIGH n.d.) show that since the tariff review of 2002, the regressivity has worsened, with the top decile receiving 15 percent of the total subsidy in 2008 (see figure 20A.8) compared with 10 percent of the total subsidy benefits in 2005 and 4.4 percent in 2003.

Social Safety Nets

Mexico's experience with government social safety net programs is considerable and has shown commendable progressive results. Experience with effective operation of social safety net programs is crucial when phasing out subsidies because poor implementation can be costly both in terms of failure to cushion the impact of higher prices on the poor as well as on government resources.

Oportunidades is Mexico's main antipoverty government program, launched in 1997. The program started as *Progresa* but changed its name in 2002. It focused on the poor and embraced an innovative approach to poverty, using conditional cash transfers that linked income transfers with preferential access to health or education services. It has been characterized by a careful system of identification and selection of beneficiaries involving the geographic selection of poor areas using a census-based marginality index; categorical criteria to identify poor households using socioeconomic survey and census data; and proxy means tests to prevent discretionary manipulation of public funds. An independent impact evaluation program has been used to improve the program's effectiveness and strengthen its political legitimacy. The program is centrally run by a federal agency that identifies and selects beneficiaries, coordinates payments and monitoring systems, and administers service delivery with ministries and federal and state agencies that are in charge of the direct provision of health and education services.

More than 70 percent of beneficiaries are regarded as extremely poor (Niño-Zarazúa 2010). *Oportunidades* has been quite successful in targeting the poor in rural communities and, in contrast to electricity subsidies, shows a high degree of progressivity (see figure 20.3). By June 2010, *Progresa-Oportunidades* covered more than 5.5 million households living in more than 103,600 localities, meaning that one out of every four families in Mexico received support from the program.

Since 2007, Mexico introduced an energy component to the program, *Oportunidades Energéticas*, to help poor households cover their energy expenses, as shown in table 20.1. The pilot effort for such a system was implemented with a small cash transfer component (relative to existing electricity subsidies) equal to a Mex$3 billion (equivalent to US$400 million) annual budget. This program did not replace or modify existing tariff-based subsidies, but it operated on a parallel basis.

In response to the 2008 financial crisis, the Mexican government introduced an additional monthly transfer of Mex$60 (equivalent to US$4.40) per household to compensate for higher energy prices. This additional transfer has been kept in place ever since.

Figure 20.3   *Oportunidades*'s Coverage of Beneficiary Households and Localities in Mexico, 1997–2010

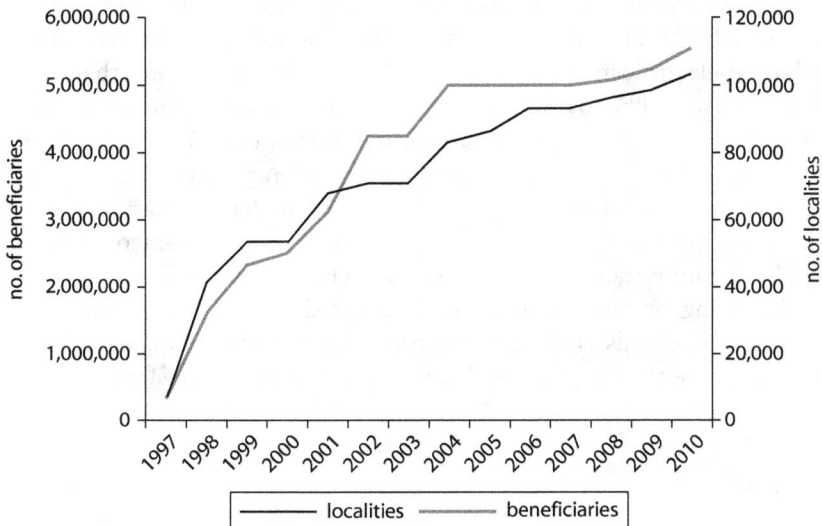

Source: Niño-Zarazúa 2010.

Table 20.1   *Oportunidades*'s Cash Transfer Components
US$

Nutrition	$19.50 per family	Cash transfers to improve family's consumption and nutrition
		Nutritional supplements for children under five and pregnant or breastfeeding women
Health		Preventive health care check-ups and health and nutrition workshops
Education	$12.60–$80.10 per student	Educational grants for children starting in third grade and up to last grade of high school and transfers for school supplies
Jóvenes con Oportunidades (since 2003)	$349.20 per graduate	Savings account for youth students who finish high school
Elderly (since 2006)	$26.20 per adult	Additional cash transfers for adults at least 70 years old who are members of urban beneficiary families
Energy Component (since 2007)	$4.80 per family	Additional economic aid for energy expenditure

Source: Yepez-García, Luna-Tovar, and Portes 2012.

Key Lessons Learned

Komives et al. (2009) simulated alternative subsidy schemes for electricity in the case of Mexico and found that a volume-differentiated tariff (as well as the use of means-tested discounts) would increase the value of targeting of the subsidy. The country's high connection rate (Mexico has already achieved electrification coverage of above 97 percent) ensures that exclusion rates would be low under any alternative schemes. Simulations show that there is practically no impact on the distribution of subsidy benefits from minor tariff "reforms" expanding their coverage to the highest-decile consumers in each tariff category (Komives et al. 2009). Phasing out subsidies is expected to have a significant impact on middle-income households. The importance of strengthening social safety nets, then, is great.

Replacing energy subsidies with targeted cash transfers for lower-income households, such as through an expanded *Oportunidades*, would help improve the targeting performance of the social transfers.

Annex 20.1 Mexico Case Study Figures

INCOME LEVEL: Upper-middle income
REGION: Latin America and the Caribbean
ENERGY NET IMPORTER/EXPORTER: Net exporter (net importer of gasoline)
SUBSIDIES: Electricity, diesel, gasoline
PHASING OUT SUBSIDIES: Ongoing

Fiscal Burden of Energy Subsidy in Mexico

Figure 20A.1   Explicit Budgetary Energy Subsidies in Mexico, 2001–10

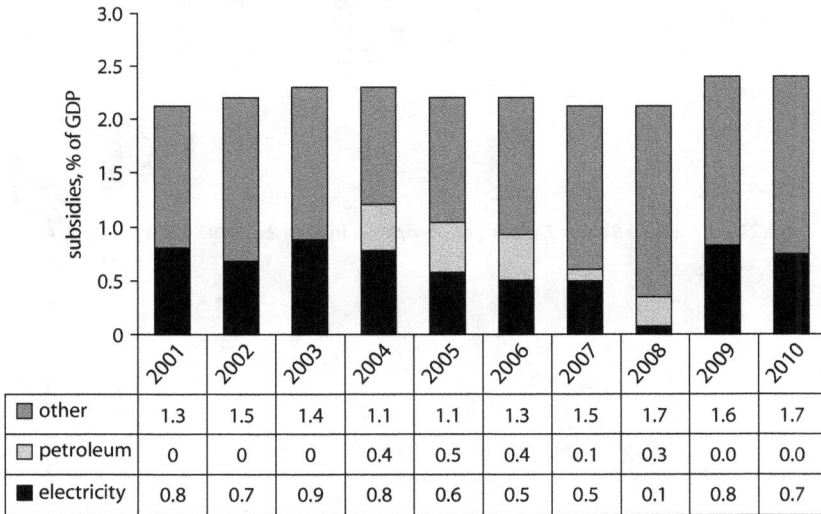

	2001	2002	2003	2004	2005	2006	2007	2008	2009	2010
other	1.3	1.5	1.4	1.1	1.1	1.3	1.5	1.7	1.6	1.7
petroleum	0	0	0	0.4	0.5	0.4	0.1	0.3	0.0	0.0
electricity	0.8	0.7	0.9	0.8	0.6	0.5	0.5	0.1	0.8	0.7

Sources: PEMEX and CFE annual reports (various years); Secretaría de Hacienda y Crédito Público.

Fuel Prices and Road Sector Consumption in Mexico

Figure 20A.2   Domestic Retail Fuel Prices in Mexico, 2002–10

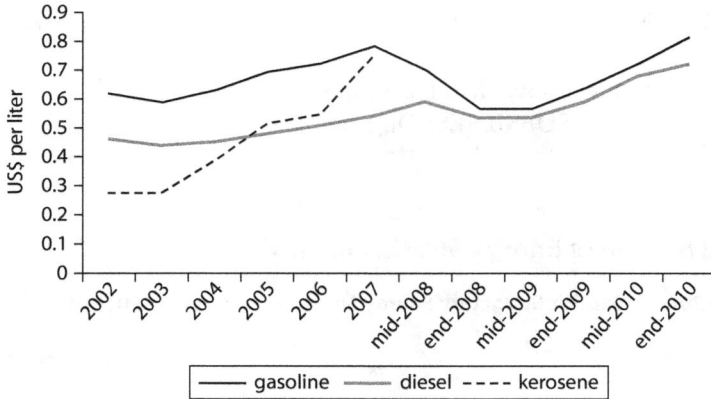

Sources: PEMEX and CFE annual reports (various years).

Figure 20A.3   Road Sector Diesel Consumption in Mexico, 1998–2008

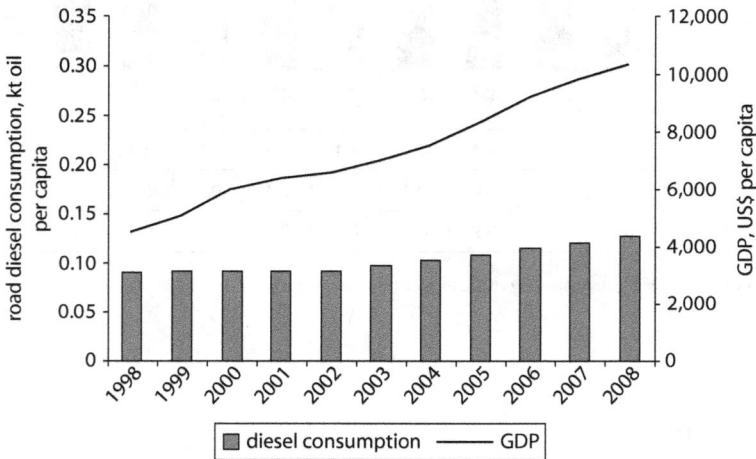

Source: World Bank, World Development Indicators.

Figure 20A.4   Road Sector Gasoline Consumption in Mexico, 1998–2008

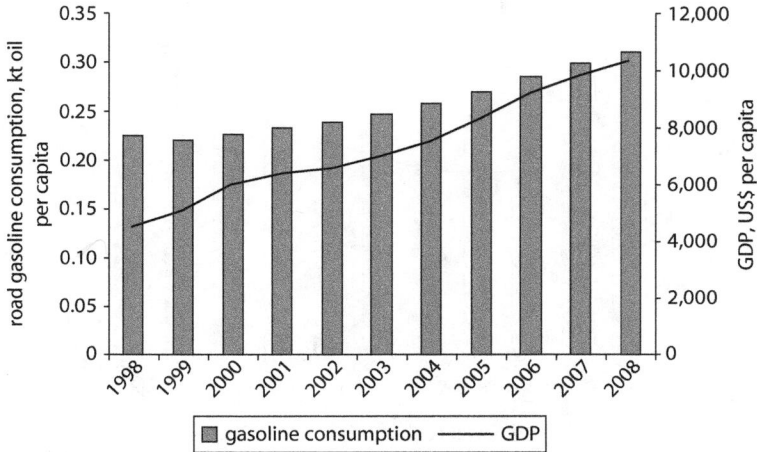

Source: World Bank, World Development Indicators.

Electricity Price and Power Consumption in Mexico

Figure 20A.5   Electricity Price in Mexico, 1998–2010

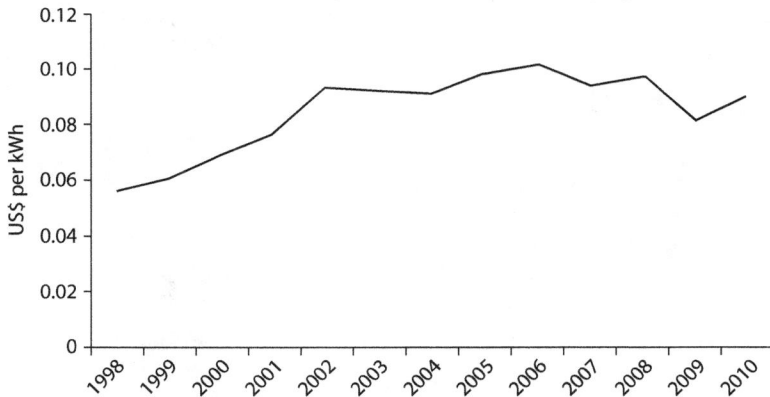

Source: CFE.
Note: kWh = kilowatt-hour.

Figure 20A.6   Power Consumption Per Capita in Mexico, 1998–2008

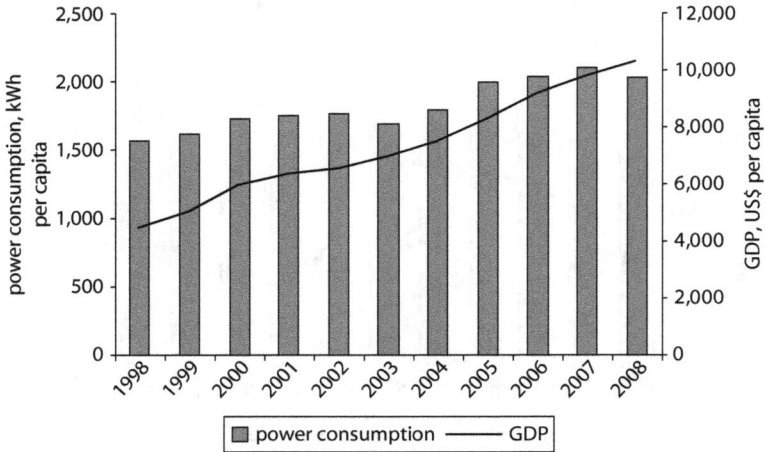

Source: World Bank, World Development Indicators.
Note: kWh = kilowatt-hour.

Poverty Impact Evidence from Household Surveys in Mexico

Figure 20A.7   Electricity Block Tariffs in Mexico, 2010

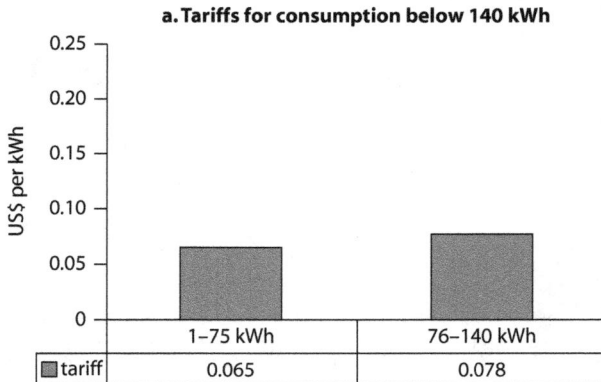

a. Tariffs for consumption below 140 kWh

	1–75 kWh	76–140 kWh
tariff	0.065	0.078

(continued next page)

Figure 20A.7   *(continued)*

b. Tariffs for all consumption levels

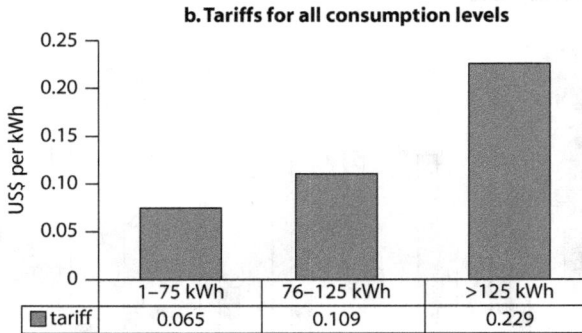

	1–75 kWh	76–125 kWh	>125 kWh
▣ tariff	0.065	0.109	0.229

Source: CFE.
Note: kWh = kilowatt-hour.

Figure 20A.8   Benefit Incidence of Energy Subsidies in Mexico, by Subsidy Type and Income Decile, 2008

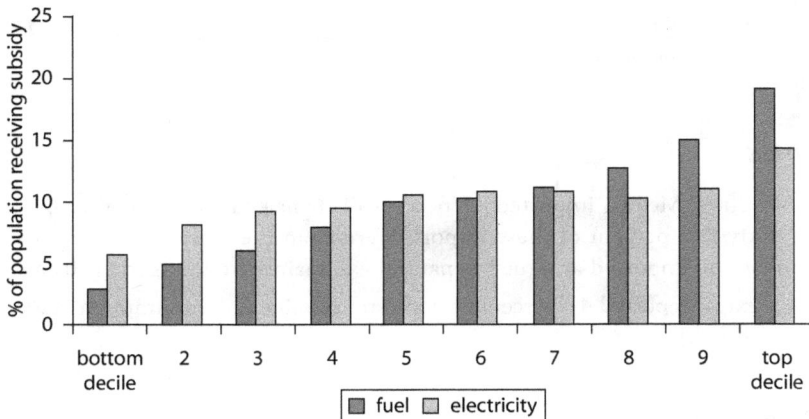

Source: ENIGH.

Figure 20A.9   Benefit Incidence of Social Expenditure in Mexico, by Benefit Type and Income Decile, 2008

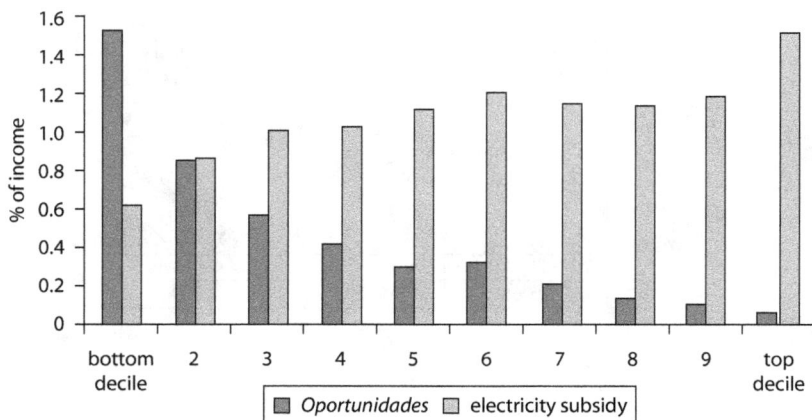

Source: ENIGH.

Table 20A.1 Concentration of Electricity Subsidy in Mexico, by Population Type, 2008

	Concentration of electricity subsidy, by group
Urban	0.14
Rural	−0.17

Source: ENIGH.

Notes

1. In 2009, Mexico imported 17 percent of its natural gas for consumption. Sixty-five percent of these imports were from the United States, and the rest was imported as liquefied natural gas, mainly from Nigeria (SIE n.d.).

2. Mexico imported 47 percent of domestic gasoline for consumption in 2010 (SIE n.d.).

References

CFE (Federal Electricity Commission). Various years. Annual Reports. CFE, Mexico City.

ENIGH (Mexican National Household Income and Expenditure Survey) (database). n.d. ENIGH (Encuesta Nacional de Ingresos y Gastos de los Hogares) database.

Komives, Kristin, Vivien Foster, Jonathan Halpern, and Quentin Wodon. 2005. *Water, Electricity, and the Poor: Who Benefits from Utility Subsidies?* Directions in Development Series. Washington, DC: World Bank.

Komives, Kristin, Todd M. Johnson, Jonathan D. Halpern, Jose Luis Aburto, and John R. Scott. 2009. "Residential Electricity Subsidies in Mexico." Working Paper 160, World Bank, Washington, DC.

Niño-Zarazúa, Miguel. 2010. "Mexico's *Progresa-Oportunidades* and the Emergence of Social Assistance in Latin America." BWPI Working Paper 142, University of Manchester, U.K.

OECD (Organisation for Economic Co-operation and Development). 2011. "Economic Surveys: Mexico." OECD, Paris.

PEMEX (*Petróleos Mexicanos*). Various years. Annual Reports. PEMEX, Mexico City.

Rosas-Flores, Jorge Alberto, and David Morillo Galvez. 2010. "What Goes Up: Recent Trends in Mexican Residential Energy Use." *Energy* 35 (6): 2596–602.

SIE (SENER [Secretaría de Energía] Energy Information System) database. n.d. SENER, Federal Government, Mexico City. http://sie.energia.gob.mx/sie/bdi.

Tepach, Reyes. 2011. "Análisis de los precios y de los subsidios a las gasolinas y el diesel en México, 2007–2011." SAE-ISS-06-11. Deputy Chamber, Economic Analysis.

Yepez-García, Rigoberto Ariel, Pedro Luna-Tovar, and Luis San Vicente Portes. 2012. "Subsidies in the Energy Sector in Mexico." Unpublished manuscript, World Bank, Washington, DC.